Sustainable Bioenergy Production

Sustainable Bioenergy Production

Editor

Veer Pratap Singh

Sustainable Bioenergy Production

Edited by **Veer Pratap Singh**

Printed in 2017

ISBN: 978-1-68117-033-6

Library of Congress Control Number: 2015931835

© 2016 by

SCITUS Academics LLC,
616, Corporate Way, Suite 2, 4766,
Valley Cottage, NY 10989

www.scitusacademics.com

Contents

Preface

Bioenergy is energy derived from biomass which includes biological material such as plants and animals, wood, waste, (hydrogen) gas, and alcohol fuels. In essence bioenergy is the utilisation of solar energy that has been bound up in biomass during the process of photosynthesis. Bioenergy is a renewable energy that can generate many additional benefits, the extent of which depends on a combination of factors including the types of feedstocks used, how they are produced and transported and the efficiency of the technologies deployed to convert them to bioenergy.

Bioenergy can provide air quality benefits where biomass residues that would otherwise be open-burnt in the field or forest, such as stubble, tree prunings or forest slash is removed and burnt in an advanced emissions controlled bioenergy plant.

Using biomass can help build resilience in agricultural, timber and food-processing industries. Bioenergy provides a use for their waste streams, can help them reduce their energy costs and potentially add a new revenue stream if they can sell biomass-derived heat and/or export 'green' electricity to the grid.

This book, Sustainable Bioenergy Production, provides comprehensive knowledge and skills for the analysis and design of sustainable biomass production, bioenergy processing, and biorefinery systems for academicians, young researchers, and professionals in the bioenergy field.

Editor

Global Bioenergy Potentials from Agricultural Land in 2050: Sensitivity to Climate Change, Diets and Yields

Helmut Haberl[a], Karl-Heinz Erb[a], Fridolin Krausmann[a], Alberte Bondeau[b], Christian Lauk[a], Christoph Müller[b], Christoph Plutzar[a], Julia K. Steinberger[a]

[a]Institute of Social Ecology, Alpen-Adria Universität Klagenfurt – Wien – Graz, Schottenfeldgasse 29, 1070 Vienna, Austria1

[b]Potsdam Institute for Climate Impact Research, PIK Potsdam, Telegraphenberg A 31, D-14473 Potsdam, Germany2

[c]Sustainability Research Institute, School of Earth and Environment, University of Leeds, Leeds LS2 9JT, United Kingdom

ABSTRACT

There is a growing recognition that the interrelations between agriculture, food, bioenergy, and climate change have to be better understood in order to derive more realistic estimates of future bioenergy potentials. This article estimates global bioenergy potentials in the year 2050, following a "food first" approach. It presents integrated food, livestock, agriculture, and bioenergy scenarios for the year 2050 based on a consistent representation of FAO projections of future agricultural development in a global biomass balance model. The model discerns 11 regions, 10 crop aggregates, 2 livestock aggregates, and 10 food aggregates. It incorporates detailed accounts of land use, global net primary production (NPP) and its human appropriation as well as socioeconomic biomass flow balances for the year 2000 that are modified according to a set of scenario assumptions to derive the biomass potential for 2050. We calculate the amount of biomass required to feed humans and livestock, considering losses between biomass supply and provision of final products. Based on this biomass balance as well as on global land-use data, we evaluate the potential to grow bioenergy crops and estimate the residue potentials from cropland (forestry is outside the scope of this study). We assess the sensitivity of the biomass potential to assumptions on diets, agricultural yields, cropland expansion and climate change. We use the dynamic global vegetation model LPJmL to evaluate possible impacts of changes in temperature, precipitation, and elevated CO_2 on agricultural yields. We find that the gross (primary) bioenergy potential ranges from 64 to 161 EJ y^{-1}, depending on climate impact, yields and diet, while the dependency on cropland expansion is weak. We conclude that food requirements for a growing world population, in particular feed required for livestock, strongly influence bioenergy potentials, and that integrated approaches are needed to optimize food and bioenergy supply.

INTRODUCTION

The surging demand of a growing and increasingly affluent world population for food, fibre, and energy is confronting the earth's terrestrial ecosystems with mounting pressures. Already today, land use is degrading the ability of ecosystems to deliver vital services to humanity [1]. Changes in the global land system are a pervasive driver of global environmental change [2] and [3]. Land-use change often leads to biodiversity loss, changes in runoff, buffering capacities of ecosystems, greenhouse gas (GHG) emissions, soil and ecosystem degradation, and other adverse effects [4]. Moreover, climate change is confronting ecosystems globally with the challenge of adapting to changes in precipitation and temperature [5], while the effects of changes in atmospheric composition, in particular increased CO_2 concentration, are currently only incompletely understood [6] and [7]. Climate change may in particular affect agro-ecosystems and is currently thought to have positive as well as negative effects on yields in different regions of the world [8].

The use of biomass for energy production as a substitute for fossil energy is often seen as an attractive option to reduce fossil-fuel dependency and help reduce greenhouse gas (GHG) emissions [9] and [10]. It has been argued that biomass combustion with consequent carbon capture and storage (CCS) on a grand scale [11], [12] and [13] might be an important option to achieve negative GHG emissions required to limit global warming to 2°Celsius until 2100, a goal thought to be required to reduce the risk of catastrophic runaway events as the earth system could reach certain "tipping points" [14] and [15]. The question of the magnitude and spatial patterns of global bioenergy potentials has therefore gained increased urgency in the last years [16], [17], [18], [19], [20] and [21].

Discussions about whether US and European biofuel policies contributed to surging prices of agricultural products and food in 2007 and 2008 [22] and [23] have drawn attention to another issue: A better understanding of the interrelations between the supply of food, fibre, and bioenergy is required in order to derive better-informed estimates of global bioenergy potentials and to forge strategies of bioenergy utilization that avoid unintended consequences such as strong increases in food prices or environmental pressures [24], [25] and [26]. Existing studies of global bioenergy potentials did so far not, or not sufficiently, consider interrelations between food and bioenergy [10], [19], [27], [28] and [29].

The interrelations between food and bioenergy depend on a host of factors, including economic factors (e.g., prices and trade), agricultural technology (e.g., crop yields, conversion efficiencies), changes in demand (e.g., diets, population numbers), as well as patterns and trajectories of global land use. This article aims to presents a first step towards the analysis of this complex system from the perspective of global socioeconomic metabolism. Studies of socioeconomic metabolism analyze the biophysical (e.g., material, energy) flows associated with human activities [30], [31], [32], [33] and [34]. This approach is based on thermodynamic principles (first and second law of thermodynamics) that allow constructing mass balances for many economic activities which complement monetary economic accounts (e.g., the System of National Accounts). Material flow analysis (MFA) can be linked with inventories of ecological material and energy flows, in particular of biomass flows, through an approach that has been called the "human appropriation of net primary production" or HANPP [35], [36] and [37]. Net primary production (NPP) denotes the amount of biomass produced by green plants through photosynthesis. HANPP records changes in the biomass balance of terrestrial ecosystems resulting from (1) human-induced changes in NPP, denoted as NPP_{LC} (NPP change resulting from land conversion) and (2) human harvest of biomass, including biomass destroyed during harvest (NPP harvested or NPP_h) [38].

Here, we use the socioeconomic metabolism approach to develop a biomass balance model to consistently link supply and demand of agricultural biomass (forestry is excluded). The model is based on a complex, data-rich representation of global supply and demand of biomass in the year 2000. We then use the model to establish a consistent biomass balance for the year 2050 based on FAO projections [39]. All biomass flows are traced from production (agriculture and grasslands) to consumption via conversion processes, in particular those related to livestock. By comparing the production potential on cropland identified by the FAO, and the production potential of grazing lands based on calculations of their primary productivity, with the biomass demand resulting from projected global food and fibre consumption, we calculate potentials to produce bioenergy on the cropland area as projected by FAO for 2050 as well as on additional cropland that could be established on grazing areas. In estimating the latter, we explicitly considered biomass demand of livestock to be satisfied from grazing land according to the projected final demand in 2050. As the model calculates mass balances for agricultural activities, it also provides data to estimate the bioenergy potential from agricultural biomass residues. We also use the biomass model to evaluate the consequences of possible effects of climate change on crop yields – as assessed by the dynamic global vegetation model LPJmL [40] – on biomass supply and bioenergy potentials.

MATERIALS AND METHODS

Definition of Study Regions and Biomass Aggregates

The regional grouping underlying this study was based on the classification of the macro-geographical (continental) regions and geographical sub-regions as defined by the United Nations Statistical Division [41]. The 11 world regions are defined in Table S1 in the supplementary online material and characterized in Table 1. Population density varies considerably between the studies regions, which is important because land availability has strong effects on land-use systems [42]. Whether a region is a net exporter or net importer of land-based products is determined by population density rather than development status [43]. Fertilizer use and livestock density are indicators of land-use intensity and differ strongly with population density as well as with per-capita income (see Table 1). The percentage of the total land area in each region used as cropland or grazing area is also indicative of land-use intensity and shows considerable differences among world regions.

Table 1: Description of the study regions in terms of area, population density and land use

	Population	Territory	Popul. density	Per-capita GDP	Livestock density	Fertilizer use	Cropland	Grazing land
Unit	[million]	[1000 km²]	[cap km⁻²]	[US$ cap⁻¹ y⁻¹]ᵃ	[LU/ha]ᵇ	[kg ha⁻¹ y⁻¹]ᶜ	[%]ᵈ	[%]ᵈ
Unit	[million]	[1000 km^2]	[cap km^{-2}]	[US\$ cap^{-1} y^{-1}]^a	[LU/ha]^b	[kg ha^{-1} y^{-1}]^c	[%]^d	[%]^d
Source	[99]	[99]	[99]	[100]	[99]	[99]	[48]	[48]
N. Africa & W. Asia	311	10,381	29.9	2753	2.43	73.3	7%	17%
Sub-Saharan Africa	650	24,291	26.8	594	2.19	10.8	7%	49%
Central Asia & Russ. Fed.	287	22,251	12.9	1762	0.89	18.7	10%	33%
E. Asia	1481	11,762	125.9	3377	4.57	229.0	14%	45%
S. Asia	1424	6787	209.8	585	9.30	98.5	35%	41%
S.-E. Asia	518	4494	115.3	935	3.15	90.8	21%	30%
N. America	314	19,600	16.0	27 818	2.00	94.8	12%	25%
Latin America, Carribean	517	20,563	25.2	2930	4.39	73.0	8%	39%
W. Europe	389	3711	104.8	23,325	6.84	185.2	24%	31%
E. & S.-E. Europe	125	1201	104.3	2401	4.47	72.3	41%	23%
Oceania & Australia	30	8559	3.5	17,223	1.56	57.7	6%	42%
World	6046	133,602	45.3	4665	3.33	88.8	12%	36%

[a]Constant 1990 US$.

[b]Livestock units (LU) per hectare of agricultural area.

[c]Kilograms of pure nitrogen (kg N) per hectare of cropland and year.

[d]Per cent of total land area.

We used the following aggregates when working with biomass production and consumption flows. We distinguished 11 food aggregates (cereals; roots and tubers; sugar crops; pulses; oil crops; vegetables and fruits; meat of ruminants (grazers); milk, butter and other dairy products; meat of pigs, poultry, and eggs; fish; other crops). We defined seven food crop aggregates (cereals; oil-bearing crops; sugar crops; pulses; roots and tubers; vegetables and fruits; others). We distinguished two groups of livestock: all animals capable of digesting roughage were aggregated into the "grazers" group (cattle, sheep, goats, etc.). All other animals (above all pigs and poultry) were included in the "non-grazers" group. Data reported in fresh weight or air-dry weight were converted into dry matter using specific data on water content according to standard tables of food and feed composition [44], [45], [46] and [47].

Data on Land Use and Global Biomass Flows in the Year 2000

Our analysis is based on a global database for the year 2000 that consistently integrates global land-use and socioeconomic data with NPP data across a range of spatial scales, from the grid level to the country level (~160 countries). Most of these data are available over the internet (http://www.uni-klu.ac.at/socec/inhalt/1088.htm). The data have been discussed extensively in previous papers [38], [48] and [49]; here we only provide a brief overview. The main strength of the database is that it covers three large domains of data that were cross-checked against one another and are consistent between scales (grid- and country-level) and domains (NPP, biomass harvest, by-products, livestock, biomass processing and use). The three main accounts are:

- A geographically explicit (5' geographic resolution, i.e. approximately 10 × 10 km at the equator) land-use dataset [48]. Cropland area and forest area are consistent with FAO data on cropland [50] and the forest resource assessments FRA and TBFRA [50] and [51] on the country level.

- A geographically explicit (5' geographic resolution) assessment of global HANPP [38]. The database includes, for each grid cell, NPP_0 (NPP of potential vegetation), NPP_{act} (NPP of the currently prevailing vegetation), and NPP_h (biomass harvested by humans, grazed by their livestock or destroyed during harvest or by human-induced fires [52]).

- A country-level assessment of socioeconomic biomass use that traces biomass flows from harvest to final consumption [49], based on FAO statistics. Flows not covered in statistics (e.g., grazing of livestock) were estimated based on country-level feed balances of all major livestock species. Livestock feed balances were cross-checked against the NPP of grazing areas [38]. Biomass harvest from cropland and permanent cultures, including primary crops, used and unused crop residues was calculated from the FAO agricultural production database [50].

Land-use data for the year 2000 are presented in Table 2. This dataset was cross-checked against statistical data and data derived from remote sensing [48]. 75.5% of the earth's land (excluding Greenland and Antarctica) is under human use which, however, ranges from very intensive to very extensive use. Approximately 1% of the land is used as infrastructure and urban area, 11.7% as cropland, 26.8% as forestry land, 36.0% as grazing land. Note that all land not classified as urban, cropland, forestry or unused land is included in the "grazing land" class, i.e. the land-use classes included in Table 2 cover the earth's entire land area. Grazing land is characterized by four quality classes (1–4, with 1 denoting the best grazing land and 4 the worst; for definitions see [48]). Land denoted as "grazing land" in our dataset therefore includes a large variety of ecosystem types: It comprises intensively cultivated meadows as well as barely productive semi-natural landscapes that often have a very high ecological value and may be used very extensively. Of the remaining 24.5%, about one half is completely unproductive, often covered by rocks and snow or deserts with an aboveground NPP below 20 g C m^{-2} y^{-1} ("non-productive land" in Table 2). The other half ("unused productive land") includes pristine forests (c.6 Mm^2; 1 Mm^2 = 10^6 m × 10^6 m = 10^{12} m^2 = 1 million square kilometers; 6 Mm^2 are approximately 4.6% of the earth's land area excluding Greenland and Antarctica), including tropical rainforests as well as all other forests with (almost) no signs of human use [53] (most of the latter in boreal regions). This category also includes rather unproductive ecosystems such as arctic or alpine tundra and grasslands.

Table 2: Land use in the 11 study regions in the year 2000. Data source [48]

	Infra-structure	Cropland	Forestry	Grazing land [1000 km²]	Non-productive land	Unused productive land	Total[a]
N. Africa and W. Asia	42	763	268	1738	7421	47	10,279
Sub-Saharan Africa	111	1781	5828	11,867	3443	945	23,975
Central Asia and Russian Fed.	189	1572	7155	6742	280	4494	20,432
E. Asia	140	1604	2121	5146	2075	448	11,533
S. Asia	113	2305	850	2554	824	024	6670
S.-E. Asia	039	931	2098	1331	0	83	4483
N. America	337	2240	4741	4473	1549	5169	18,508
Latin America & the Carribean	64	1685	8733	7932	256	1624	20,295
W. Europe	198	862	1318	1130	11	136	3655
E. & S.-E. Europe	103	941	630	482	0	2	2158
Oceania and Australia	23	540	1216	3484	305	2817	8385
World	1360	15,225	34,958	46,881	16,163	15,788	130,375

[a]The total refers to territorial surface area without inland water bodies.

Matching Supply And Demand: The Biomass Balance Model

The biomass balance model (for reference see [54]) allows to calculate scenarios of the supply and demand of biomass in 2050, based on a consistent set of assumptions discussed in section 2.4. The databases described in section 2.2 were used to construct a model of biomass flows in the year 2000 in which the demand for final products is matched with gross agricultural production and land-use data (Fig. 1). We used factors derived from data for 2000 to characterize the conversion of biomass from primary harvest to final products (food and fibre), in particular through the livestock system. The model consists of two process pathways, a food crop path for the demand for cereals, roots and tubers, sugar crops, pulses, oil crops, vegetables and fruits, and other crops, and also for the demand for pig meat, poultry, eggs, and fish from aquaculture ("non-grazers"), and a roughage path for the demand for products derived from grazers (meat, milk, butter, and other dairy products).

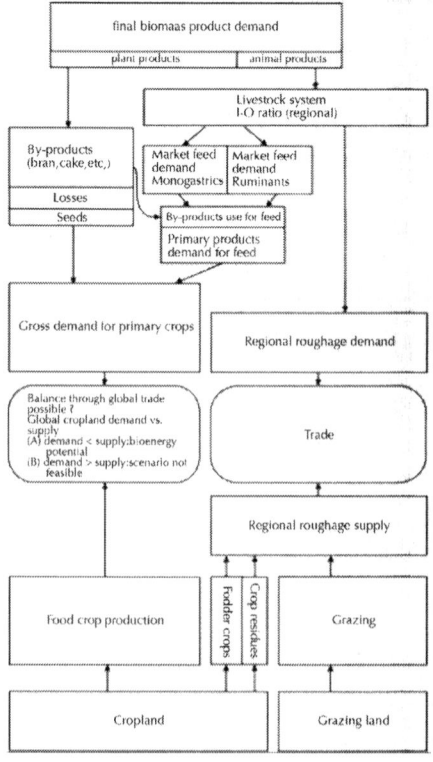

Figure 1: Flow chart of the biomass-balance model used to integrate supply and demand of biomass. For reference see [54].

In the food crop path, the regional demand for final biomass products (e.g. flour, vegetable oils, refined sugar) is converted to the amount of gross primary crop demand (i.e., primary products such as cereals, oil crops, sugar crops, etc.). Using global factors derived from the databases described in section 2.2, the by-products accruing from the production of final products (e.g. brans in flour production from cereals, oil-cakes in vegetable oil production from oil-bearing crops), seed requirements and the losses in the agricultural system are calculated (Fig. 1). Non-grazers (pigs, poultry) are dealt with in the food crop path as well, because they are fed (mainly) from primary or secondary cropland products. For the demand for final products (i.e. meat from pigs and poultry, eggs, and fish from aquaculture), the market feed requirement (e.g., brans, oil cakes, cereals) is calculated by applying regional input-output ratios of the monogastric livestock systems [49] and [55]. The resulting amount of market feed demand of non-grazers is added to the market feed demand of grazers calculated in the roughage path (see next paragraph), resulting in total regional market feed demand. This is then balanced with the regional supply of market feed from food processing and industrial processing of cereals, oil-bearing crops, and sugar crops; i.e., the supply of brans, oil-cakes, molasse, and bagasse. Usage factors for these categories were derived from the 2000 database and used to calculate the amount of market feed fed to animals. From the difference between markets feed demand and the amount of by-products from processing fed to animals, the additional demand for feed grain (cereals) is calculated and added to the regional demand for cereals, taking seed demand and losses into account.

The roughage pathway refers to the demand for ruminant meat and milk, i.e. to the grazing livestock system. The grazing livestock system is characterized by a demand for market feed and a demand for non-market feed (roughage demand; i.e., the sum of fodder, crop residues fed to grazers, and grazing). The amount of feed demand per unit of output (meat or milk) varies between world regions by factors of up to 10, due to the differences in animal husbandry systems [49]. These factors depend particularly on the regional share of subsistence livestock systems (with high input-output ratios for roughage and low input-output ratios for market feed) and industrial meat and milk production (with the opposite patterns and a much higher overall efficiency due to the higher nutritional value of market feed and a production system optimized for high outputs). We calculated the regional production of ruminant meat and milk (and subsequently regional feed demand) as a function of regional roughage supply. Crop residue flows and the fractions used as feed were derived from the databases for 2000 using data on harvest indices (the ratio of grain to total plant biomass) and the usage of harvest residues as well as data on the fraction of available crop residues used for feed [38], [49] and [56]. Fodder supply is given in FAO statistics and was converted to dry matter using standard tables, as described in section 2.1. The amount of grazed biomass was calculated from grazing land statistics [48], the actual NPP of grazing systems, and grazing intensity, i.e. the ratio of grazed biomass and actual NPP in a region [38]. The amount of total regional roughage supply was used to calculate the amount of ruminant meat and milk production in each region based on the input-output ratio of the livestock systems. From regional ruminant meat and milk production, the regional market feed demand of ruminants was derived and added to the total market feed demand.

The gap between regional supply and demand in 2000, for meat as well as for cropland products, was assumed to be balanced by international trade: for example, regions where the demand for primary products (e.g., cereals) exceeded regional supply were assumed to import; regions, where biomass supply was larger than regional demand were assumed to export. Overall, the level of uncertainty of the biomass flow model is at a satisfactory level: extrapolated global demand for gross primary crops is at 98% of the 2000 cropland production, and modelled grazing is at 99% of the grazing amount from the HANPP assessment in the year 2000 [38]. Discrepancies result from the usage of global average factors. In order to use the model to calculate bioenergy potentials for the year 2050, we modified the original model for the year 2000 as described in section 2.4.

Assumptions for Changes until 2050 Compared To 2000

With respect to population growth, we used the UN medium variant in which world population is forecast to be 9.16 billion in 2050 [57]. Total food demand was derived from forecast population numbers assuming "business-as-usual" changes in regional diets which we derived as follows. For the year 2000 we used data on food supply as compiled by the FAO [58], averaged over the period 1999–2001 in order to avoid climate or other fluctuations, and aggregated to the food categories described in section 2.1. By 2050, every region was projected to attain the diet level of the country which was "richest" (in terms of food intake) in 2000 in the respective region. The composition of the richest country's diet was adapted to the regional pattern in order to maintain appropriate fractions (for instance for pork meat in the Islamic countries of North Africa and Western Asia). The diet projected for 2050 is compared to that of 2000 in Table 3. This business-as-usual (BAU) scenario is quite similar to the business-as-usual demand growth scenarios of the FAO for 2050 [39], despite the difference in methodology [59].

In order to test the sensitivity of the bioenergy potential in 2050 to diets, we performed an alternative model run, assuming a global food supply of 11.72 MJ cap^{-1} d^{-1} (i.e. the current global average) with only 7–10% of the calorific energy animal products (see Table 3). While this "fair and frugal" diet was designed to be nutritionally sufficient in terms of calorie and protein supply, it would require equitable distribution of food in order to avoid malnutrition and imply a quite significant reduction in terms of calorie supply as well as consumption of animal products in some parts of the world. It is included here to demonstrate the dependency of bioenergy potentials on future changes in diets.

Table 3: Food supply in 2000 and two assumptions for the year 2050: A "business-as-usual" forecast (BAU) as well as a "fair and frugal" diet ("fair") assuming a switch to equitable food distribution and less meat consumption. Absolute numbers are kilocalories per capita per day [MJ cap^{-1} d^{-1}]

	Total food supply 2000	Share of animal products 2000	Total food BAU 2050	Change in total, BAU 2050/2000[MJ cap^{-1} d^{-1}] or per cent [%]	Share animal products BAU	Total food "fair" 2050	Change in total, "fair" 2050/2000	Share animal products "fair"
N. Africa and W. Asia	12.38	10%	13.37	8%	12%	11.72	-5%	8%
Sub-Saharan Africa	9.41	7%	11.73	25%	8%	11.72	25%	8%
Central Asia, Russ. Fed.	11.66	22%	12.87	10%	23%	11.72	1%	8%
E. Asia	12.29	19%	13.16	7%	21%	11.72	-5%	8%
S. Asia	10.15	9%	11.52	13%	13%	11.72	15%	10%
S. -E. Asia	11.21	8%	11.98	7%	11%	11.72	5%	8%
N. America	15.69	27%	15.70	0%	27%	11.72	-25%	7%
Latin America, Carrib.	11.87	20%	12.82	8%	21%	11.72	-1%	8%
W. Europe	14.36	31%	14.75	3%	32%	11.72	-18%	7%
E. & S.-E. Europe	12.86	25%	13.62	6%	27%	11.72	-9%	9%
Oceania and Australia	12.63	28%	13.46	7%	29%	11.72	-7%	7%
World	11.67	16%	12.53	7%	16%	11.72	0%	8%

We used the UN population forecast [57] to derive an estimate of the additional area needed for urban areas and infrastructure as follows. We assumed that rural infrastructure areas are mostly driven by the need to transport agricultural inputs and produce and by the need to house agricultural population and machinery. We therefore calculated the area of rural infrastructure as a percentage of cropland area in each region, using factors derived from prior work [48]. Urban areas are much smaller than rural infrastructure. We estimated urban areas in 2050 by assuming that the per-capita amount of urban area would stay constant from 2000 to 2050. Globally, urban population is forecast to increase from 2.84 to 6.37 billion [57]. For East and South-East Europe, the UN forecasts a shrinking urban population; in this region we kept the urban areas constant. We are aware that such simple assumptions can only serve to derive first-order approximations that might be too low; that is, the results are likely to be conservative. According to our calculation, urban areas grow from 279,180 km^2 to 532,880 km^2. This is not much when compared with existing cropland areas (Table 2), so the ensuing errors introduced by our estimation method will also be small.

We used FAO forecasts [39] and [60] to derive estimates for cropland area change and crop yields until 2050 (for reference see [54]). The FAO provides projections of crop production and its drivers (yields, area, and cropping intensity) for selected important food crops (cereals, oil crops, sugar crops) for industrialized countries and five regional groups of developing countries [39] and [60]. FAO projections are not based on a formal model, but use expert judgements, mostly of FAO in-house experts, to derive estimates of demand for food, feed, non-food uses, seeds and wastes as well as regionally specific projections of yields and cropped areas. Balances between supply and demand are closed using so-called "supply-utilization accounts" (SUA's). The projections have to fulfil consistency criteria and are improved in an iterative process that involves several stages of revision, ensuring that sectoral and regional knowledge can be incorporated [60].

When these were available, we applied annual growth rates of crop production and its drivers (area, yield, cropping intensity) as reported by the FAO to our data [48] and [19] to derive total production volumes and area changes for crops and regions explicitly covered in the relevant reports [39] and [60] (yields were cross-checked, and slightly modified, using GAEZ data [61]). In order to avoid complications arising from working with "harvest yields" (i.e., yields per harvest event; areas with multicropping are counted each time they are harvested, fallow is omitted), we use the concept of "land-use yields" (derived by dividing the total amount of crops produced per unit of cropland area, including fallow). Land-use yields are calculated by multiplying harvest yields and cropping intensity; i.e., the number of harvests per year. Results are shown in Fig. 2. The FAO does not report projections for fodder crop production. To fill this gap, we assumed that the share of fodder crops to the overall area of arable land remains constant and that the yields of fodder crops grow with the same rate as the aggregate "other crops".

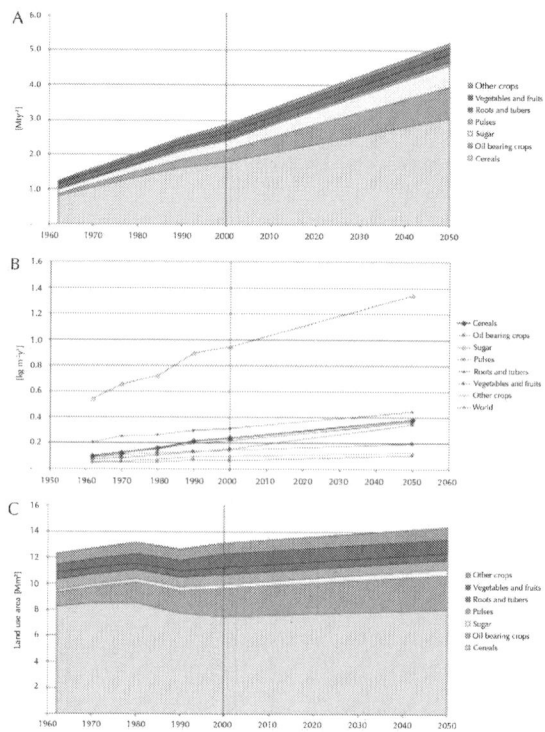

Figure 2: Cropland production scenario until 2050. Trajectory of (A) production, (B) land-use yields (= harvest yield times cropping intensity) and (C) cropland area 1960–2050 of food crops, break-down to major crop groups. Material flow data are reported in metric tons of dry-matter biomass. For sources and details, see text.

The results are plausible compared with current crop yields at the national scale [50] and alternative yield forecasts [62]. Our assumptions deviate from FAO projections only marginally, especially when compared to the level of uncertainty in such a projection. Overall, we assumed that cropland area will grow by 9% (Table 4) and yields by 54% (Fig. 2). Our assumptions are in line with other studies: IIASA scenarios suggest that global cropland area will grow by +6% in scenario B1, +9% in Scenario B2 and +12% in scenario A1 until 2050 (http://www.iiasa.ac.at/Research/GGI/). Most global agricultural scenarios assume that growth in agricultural production will depend mostly on increases of yields and only to a smaller extent on a growth of cropland areas [63] and [64].

Table 4: Cropland areas and changes in 2000 and 2050, according to our re-calculation of the FAO scenario "World agriculture towards 2030/50" (FAO/BAU) and an alternative "massive expansion" assumption

	Cropland in the year 2000	Cropland in 2050 FAO/BAU		Cropland in year 2050 massive expansion	
	[1000 km²]	[1000 km²]	[change]	[1000 km²]	[change]
Northern Africa and Western Asia	763	819	+7.2%	874	+14.5%
Sub-Saharan Africa	1781	2283	+28.2%	2785	+56.3%
Central Asia and Russian Federation	1572	1635	+4.0%	1699	+8.1%
Eastern Asia	1604	1694	+5.7%	1785	+11.3%
Southern Asia	2305	2428	+5.3%	2550	+10.6%
South-Eastern Asia	931	930	−0.1%	931	0.0%
Northern America	2240	2335	+4.3%	2430	+8.5%
Latin America & the Carribean	1685	2037	+20.9%	2388	+41.7%
Western Europe	862	880	+2.1%	899	+4.2%
Eastern & South-Eastern Europe	941	890	−5.4%	941	0.0%
Oceania and Australia	540	696	+28.8%	851	+57.7%
World	15,225	16,627	+9.2%	18,134	+19.1%

In order to test the sensitivity of our calculations to assumptions on yields and cropland expansion, we also ran the model with the following assumptions: According to the scenario report of the "Millennium Ecosystem Assessment" (MEA) [1], the "TechnoGarden" scenario is comparable with FAO forecasts. The highest and the lowest yield scenarios in MEA span a range of +9% to −19% around that scenario; we used this range for our sensitivity analysis. With respect to cropland area, we also ran a scenario in which growth of cropland area was doubled in all regions and held constant in all regions where FAO forecasts shrinking cropland areas. In this expansion scenario, cropland area is assumed to grow by +19% until 2050 compared to the year 2000 (Table 4).

As this study focuses on agriculture and excludes forestry, we made the conservative assumption that growth in cropland and urban/infrastructure area reduces the area of grazing lands only, while forest areas remain constant. We assumed that the area expansion of cropland and infrastructure consumes the best grazing areas, i.e. that of class 1 and in regions where sufficient grazing land of that quality class is available, and class 2 where this is not the case (i.e. North Africa and Western Asia). The biomass-balance model calculates grazing intensity on grazing land (i.e. the ratio of biomass grazed to NPP_{act} on grazing land) as discussed in section 2.2 (the allocation to grazing land quality classes is described in [38]). Our pattern of cropland expansion (Table 3) is comparable to other studies on global cropland potentials [65] and cropland suitability maps [66].

Based on statistical data reported by the FAO and standardized according to methods described elsewhere [49], we derived trajectories of the input-output ratios of livestock for the time period from 1961 to 2000 at the regional level which we projected until 2050 based on data on feeding efficiencies of different livestock rearing systems (see [54]). These input-output ratios reflect an assumed reduction of the respective regional subsistence fractions by 50% in favour of industrial, indoor-housed, or extensive, market-oriented production systems, depending on area availability (Table 1). Data for 1961–2000 and our projection for 2050 are shown in Fig. 3.

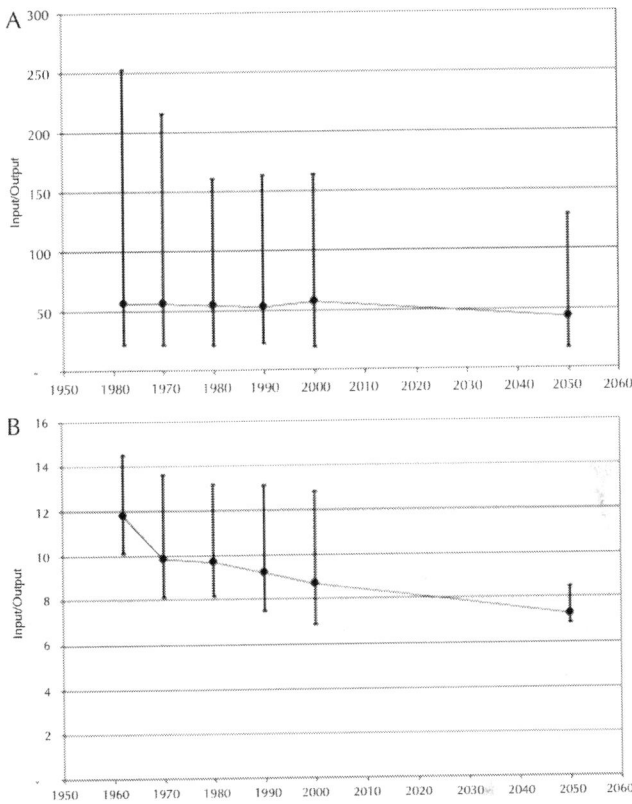

Figure 3: Development of livestock input-output ratios 1962–2050. Feed demand of A) Grazers (cattle and buffalo, sheep, goats), B) Non-grazers (pigs, poultry). These input-output ratios refer to the overall regional feed demand of the entire livestock population in each region ("top down"). Dots indicate the weighted global average, whiskers the ranges between regions. For details, see text.

Calculation of Bioenergy Potentials

We calculated bioenergy potentials by distinguishing three fundamentally different production pathways: (1) bioenergy crops on cropland, (2) bioenergy crops on other lands (i.e. grazing land according to the land-use dataset used in this study), and (3) residue potentials on cropland. We calculated gross potentials for bioenergy supply by assuming that the entire aboveground NPP of bioenergy crops can be used to produce bioenergy, assuming a gross calorific value of dry-matter biomass of 18.5 MJ kg^{-1}[67]. The calculation did not take conversion or production losses into account.

In order to calculate the area available for producing bioenergy on cropland, we subtracted the area required for food, feed, and fibre calculated with the biomass-balance model described in section 2.3 in each region from each region's cropland area (section 2.4). We calculated the bioenergy potential by assuming that the NPP of bioenergy crops is equal to potential NPP [68] and [69] and that the entire aboveground biomass can be harvested and used to produce bioenergy. Data on potential NPP (NPP_0) were taken from previous work [38].

To calculate the potential to grow bioenergy crops on other land (i.e. grazing areas, see section 2.2), we assume that grazing land in the quality class 1 is also suitable for producing of bioenergy crops such as switchgrass (*Panicum virgatum*), *Miscanthus* sp, short-rotation coppice or similar bioenergy crops. This seems justified as a cross-check of the regional distribution of grazing areas in quality class 1 with the regional distribution of cropland potentials/suitability [65] and [66] revealed that regions with large cropland potentials also have large areas of high-quality grazing land and vice versa. We assume that grazing on land in grazing quality class 1 can be intensified, assuming an increase of the exploitation rate of NPP_{act} to a maximum of 67% in developing and 75% in industrialized regions. This allows using a significant fraction of the area in grazing land of quality class 1 for bioenergy crops without reducing regional roughage supply. On the area that becomes available for bioenergy crops through intensification, the bioenergy potential is estimated to be equal to aboveground NPP_{act} (taken from [38]); that is, we assume that bioenergy crops produce the same amount of aboveground biomass per year as the current vegetation [69] and [68].

The energy potential from unused residues on cropland was calculated by applying harvest indices and usage factors described in section 2.3. Crop residues are used as feed and for bedding. The bedding requirement was estimated by calculating the amount of manure produced by livestock and applying factors to estimate bedding demand from indoor manure production derived from [49]. We assumed that 50% of the remaining residues are required to maintain soil fertility and are therefore not available to produce bioenergy [16].

We are aware that this is a crude assumption and that higher or lower shares of the residues might be required to maintain soil fertility in different regions, depending on soil and climate conditions [70].

Modelling of Climate Change Effects with LPJmL

We employed the LPJmL model [40] to estimate the effects of changes in temperature, precipitation and CO_2 fertilization on yields of major crops globally at a spatial resolution of 0.5° × 0.5°. Yield calculations were based on process-based simulations of 11 agricultural crops in a mechanistic coupled plant growth and water-balance model (for reference, see [40]).

We calculated percent changes in agricultural productivity between two 10-year periods: 1996–2005 and 2046–2055, representing the average productivity of the years 2000 and 2050. Management intensity was calibrated to match national yield levels as reported by FAO statistics for the 1990s [71]. National and regional agricultural productivities were based on calorie- and area-weighted mean crop productivity of wheat, rice, maize, millet, field pea, sugar beet, sweet potato, soybean, groundnut, sunflower, and rapeseed. LPJmL simulations were used only to estimate the possible magnitude of the climate-change effect on agricultural yields. In these simulations we assumed constant management intensities and cropping patterns as of the year 2000. Changes in management, breeding and cropping area were covered by other data and assumptions as described in sections 2.3 and 2.4. We did not consider feedbacks between climate change, CO_2 fertilization, and management. Still, our results provide a sound estimate of possible impacts of climate change on agricultural yields with and without CO_2 fertilization effects.

We assumed three different emission scenarios from the Special Report on Emission Scenarios (SRES): A1b, A2, B1 [72]. Each emission scenario was implemented in five different general circulation models (GCMs): CCSM3 [73], ECHAM5 [74], ECHO-G [75], GFDL [76], and HadCM3 [77]. Climate data for these GCM-projections were generated by downscaling the change rates of monthly mean temperatures and monthly precipitation to 0.5° resolution by bi-linear interpolation and superimposing these monthly climate anomalies (absolute for temperature, relative for precipitation and cloudiness) on the 1961–1990 average of the observed climate [78] and [79]. Since there is no information about the number of wet days in the future, we kept these constant after 2003 at the 30-year average of 1971–2000.

Considerable uncertainty exists how CO_2 fertilization might influence future crop yields. This is due to both modelling uncertainties and to the fact that it seems likely that there are interrelations between management (e.g., nutrient and water availability) and the CO_2 fertilization effect. To assess the range of CO_2 fertilization uncertainty [6] and [7], each of the 15 scenarios was calculated twice: first, taking into account full CO_2 fertilization effects according to the prescribed SRES atmospheric CO_2 concentrations, and second, keeping atmospheric CO_2 concentrations constant at 370 ppm after 2000. In the latter case, yield changes are only driven by the modelled changes in precipitation and temperature (and the limited adaptation of management as described below), whereas in the first case the full effect of changes in temperature, precipitation, and atmospheric CO_2 levels is taken into account. Relative management levels were kept static, but sowing dates were assumed to be adapted to climate change as described by [40] and for wheat, maize, sunflower, and rapeseed (but not for all other crops) we also assumed adaptation in selecting suitable varieties.

Yield data were originally calculated at a spatial resolution of 0.5° × 0.5° and then aggregated to country-level change rates. We then calculated the arithmetic mean of the change rates in all 15 scenarios with and without CO_2 fertilization effect. These country-level results were then used to calculate the area-weighted average deviation of the crop yields in each region from the yield levels projected by the FAO.

RESULTS

Our estimates of changes in crop yields resulting from climate change are presented as region-specific percent change rates in Table 5. We found that crop yields increase (compared to the BAU scenario) in all 11 regions if full CO_2 fertilization is assumed, but the growth varies considerably between regions from +0.74% to +28.22% (area-weighted average: +14.76%). If the CO_2 fertilization effect is switched off, however, we find considerable losses (compared to the BAU scenario) of up to −16.02% in most regions, although some regions might still benefit (up to +5.12%); the average (area-weighted) loss of cropland yields was −7.06%.

Table 5: Modeled climate impact on cropland yields in 2050 with and without CO_2 fertilization

	Mean yield change under climate change 2050	
	with CO_2 fertilization	**without CO_2 fertilization**
Northern Africa and Western Asia	+ 4.44%	−8.65%
Sub-Saharan Africa	+8.46%	−6.17%
Central Asia and Russian Federation	+24.91%	+5.12%
Eastern Asia	+11.96%	−3.90%
Southern Asia	+18.45%	−15.61%
South-Eastern Asia	+28.22%	−15.83%
Northern America	+12.45%	−6.25%
Latin America & the Carribean	+12.39%	−7.02%
Western Europe	+16.42%	+ 2.04%
Eastern & South-Eastern Europe	+19.08%	−0.66%
Oceania and Australia	+0.74%	−16.02%

The calculated global bioenergy potential in the absence of climate change ("business-as-usual" or BAU) is reported in Table 6. We found that the global aggregate primary bioenergy potential in the year 2050 without climate change amounts to 104.7 EJ y^{-1}. More than half of that potential comes from primary crops on other (grazing) land, i.e. from the intensification of land use on the best available grazing areas. Residues and primary crops on cropland assumed to exist in 2050 according to FAO projections (see Table 4) contribute less than 50%. Almost half of the potential comes from only two regions, namely Sub-Saharan Africa and Latin America and the Caribbean. Two other regions, Northern America and South-Eastern Asia contribute another quarter, whereas the other regions are only minor contributors.

Table 6: Modeled bioenergy potentials in the "business-as-usual" (BAU) scenario in the year 2050 (excluding climate change)

	Primary crops on cropland [EJ y^{-1}]	Residues on cropland [EJ y^{-1}]	Primary crops on grazing land [EJ y^{-1}]	Total [EJ y^{-1}]
Northern Africa and Western Asia	0.02	1.08	0.00	1.11
Sub-Saharan Africa	0.75	2.19	20.50	23.44
Central Asia and Russian Federation	0.88	1.08	5.95	7.91
Eastern Asia	0.48	5.06	1.30	6.83
Southern Asia	0.65	2.29	0.00	2.94
South-Eastern Asia	1.94	2.75	6.43	11.11
Northern America	5.91	5.97	3.67	15.55
Latin America & the Carribean	4.91	2.39	16.69	23.99
Western Europe	0.34	2.57	0.67	3.59
Eastern & South-Eastern Europe	1.85	1.91	2.58	6.34
Oceania and Australia	0.24	0.35	1.30	1.89
World	17.97	27.63	59.10	104.70

Climate change could result in changes in cropland yields (Table 5) and in the productivity of grazing areas that would have a considerable effect on the modeled bioenergy potential, as shown in Fig. 4a: If the CO_2 fertilization effect, as modeled by LPJmL, is fully effective, the bioenergy potential might rise by up to 45% to 151.7 EJ y^{-1}, whereas it would decrease by 1690 to 87.5 EJ y^{-1} if CO_2 fertilization is assumed to be completely ineffective. Fig. 4b shows that this is only partly a result of increased yields on areas used for growing bioenergy: Growth in yields compared to BAU makes more area available for growing bioenergy, while any reduction in cropland yields results in less area availability. This implies that the global bioenergy potential on cropland and grazing areas is highly dependent on the (uncertain) effect of climate change on future global yields on agricultural areas. We found that the potential of primary bioenergy on cropland is most sensitive to climate change, whereas the potential on grazing areas and the residue potential is less affected by climate change. Note, however, that the distinction between primary bioenergy crops on cropland and grazing land is to some extent arbitrary in the sense that assuming a larger extension of cropland until 2050 increases the potential for primary bioenergy crops on cropland at the expense of the potential for primary bioenergy crops on grazing land, under *ceteris paribus* conditions (see below).

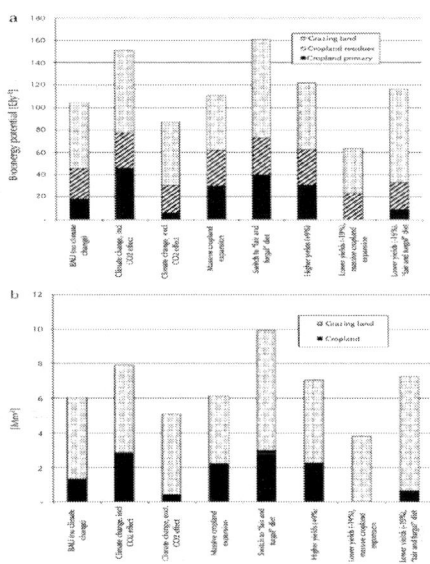

Figure 4: Comparison of the bioenergy potential and area used in the "business-as-usual" (BAU) scenario compared to variants in which one or two parameters were modified (all other assumptions are identical to BAU). (a) Bioenergy potential from cropland, residues and grazing land; (b) area used to grow plants designated for bioenergy use: Cropland areas and grazing areas converted to bioenergy plantations.

Fig. 4 also shows that the higher growth in cropland areas assumed in the "massive expansion" variant would have a small effect on the bioenergy potential (which would rise by about 6% to 110.5 EJ y^{-1} compared to BAU). The reason is the following: Cropland expansion would allow to produce more bioenergy on cropland, but less bioenergy on grazing land, as the expansion of cropland would reduce the area of grazing land and therefore the potential to grow bioenergy there without jeopardizing feed demand.

A switch to a "fair and frugal" diet would have a major impact on the bioenergy potential, however, which might be as high as 160.8 EJ y^{-1} (+54%) under these conditions. If we assume higher yields (and a BAU diet), the bioenergy potential rises to 121.6 EJ y^{-1} (+16%). If yields were to be 19% lower than assumed in the FAO/BAU scenario, it would not be possible to produce enough biomass for the BAU diet. We therefore modelled two alternative scenarios, one that combines lower yields with a massive expansion of croplands, and one that combines lower yields with a "fair and frugal" diet. In the first case, the available cropland area is just about sufficient to produce enough food, so bioenergy could in that case only be derived from residues and grazing areas, and the potential drops to 63.6 EJ y^{-1} (−39%). In the second case, the bioenergy potential is even higher than under BAU conditions and amounts to 116.5 EJ y^{-1} (+11%).

DISCUSSION AND CONCLUSIONS

How Realistic is the FAO Forecast Underlying this Study?

The results of this study are based on the FAO projections which describe a world of improved food supply and rapid agricultural intensification. Overall production on cropland increases by 68% (dry matter); maximum increases are forecast for Sub-saharan Africa (+154%) and for Latin America (+121%). In these regions, the FAO also assumes a considerable expansion of cropland, in line with studies of cropland potentials/suitability [65] and [66]. Note, however, that such area potential studies have been criticized [80] and that it might be difficult to cultivate the soils prevailing in these regions with currently prevailing technologies [64] and [81].

The largest part of the growth in total production is due to growing yields, which were assumed to increase by 54% on average for all cropland. In particular, in Western Europe and North America, cropland yields reach very high levels. It is difficult to judge whether such yield gains can be realized. It has been argued that in some regions, most options to achieve yield gains have already been implemented and yields are therefore approaching physiological limits, that the best agricultural lands are already in use and area expansions may result in the use of less well-suited land, and that soil erosion and depletion of nutrient stocks in soils may pose challenges for future yield growth [82], [83] and [84]. However, improved management could help to sustain yield growth; e.g., due to improved stress tolerance, avoidance of nutrient and water shortages, or improvements in pest control. Substantial investments will be indispensable for maintaining growth in crop yields [85]. Lower rates of yield growth would result in a lower bioenergy potential, as shown in Fig. 4, while higher yields would help to increase the bioenergy potential. Achieving high yield gains might, however, result in substantial detrimental environmental impacts such as soil degradation, air and water pollution, biodiversity loss and others [64]. Judging what amount of agricultural intensification might be justified in order to increase the bioenergy potential is a complex issue that is beyond the scope of this article. Answers to this question will, among others, also depend on future development in agricultural technology [64].

Our alternative diet scenario has also shown that changes in diets compared to often-expected trajectories (growth in calorie supply and more animal products) might result in considerably higher bioenergy potentials. It should be noted, however, that the "fair and frugal" diet modelled here might be considered to be near to the lower boundary of the possibility space for that parameter, while food demand might also be thought to grow more strongly than modelled here (or by the FAO), as the global average 2050 in the BAU scenario is well below levels of food and animal product supply enjoyed today in regions such as the US and Western Europe [54] (see Table 3).

Uncertainties Regarding Climate Change Impacts

The climate change effect on crop yields is highly uncertain. Depending on climate scenario (not shown) and the assumptions on the effectiveness of CO_2 fertilization, most regions may experience significant decreases in crop yields as well as significant increases. The most important factor is the uncertainty in CO_2 fertilization which was explicitly analyzed here. This effect can, in principle, increase crop yields considerably due to enhanced carbon assimilation rates as well as improved water-use efficiency. Whether or not farmers will be able to attain increased crop yields under elevated atmospheric CO_2 concentrations will depend on the availability of additional inputs, especially nitrogen [86]. Increased carbon assimilation rates can only be converted into productive plant tissue or the harvested storage organs if sufficient nutrients are available to sustain additional growth. Where plant growth is constrained by nutrient limitations, additional growth is limited. On top of that, there is some likelihood that the quality of agricultural products decreases under increased CO_2 fertilization, as e.g. the protein content diminishes [87]. There is also evidence that crops grown under elevated CO_2 concentrations might be more susceptible to insect pests [88].

A positive climate-change effect on crop yields may be expected in regions currently constrained by too low temperatures, as in the northern high latitudes and in mountainous regions. Here, all 30 model runs uniformly indicate increases in crop yields by 2050. By contrast, there is hardly any location where all model runs uniformly indicate decreases in crop yields if CO_2 fertilization is assumed to occur. If the CO_2 fertilization is switched off, however, many regions, especially tropical croplands are uniformly projected to experience decreases in crop yields in all 15 climate scenarios. It has to be noted that the beneficial effects of CO_2 fertilization are subject to heavy debate [6] and [7]. Results presented here only indicate the order of magnitude of climate-related impacts on crop yields. Besides uncertainties in future development of drivers (climate change, CO_2 fertilization effect, management, technological change), modelling of crop yields at large scales adds to the overall uncertainty as many processes are necessarily implemented only in a simplified manner. If farmers have access to a broad selection of crop varieties, they are likely to select varieties most suited for the local growing conditions, which could not be fully considered here.

Interpretation of Bioenergy Potential Calculations

When interpreting the calculated bioenergy potentials it is essential to keep in mind that these are gross potentials for bioenergy supply; that is, the gross calorific value (GCV) of the entire aboveground plant material assumed to be available for as feedstock for bioenergy production (section 2.5). If one assumes that the plant material is directly used for combustion for heat or combined heat and power (CHP) without much (or any) conversion, this is a reasonable approximation of the primary energy available. The production of liquid biofuels with current (first-generation) technologies, however, can only convert parts of the plants into fuels and entails substantial losses due to the conversion process. On the other hand, first-generation biofuel production would also deliver feed which is not considered in our biomass balances. A considerable fraction of the bioenergy potential calculated here would not be suitable for this utilization pathway, for example the residue potential and an unknown part of the potential on grazing areas. Even in areas where first-generation biofuel production would be possible, the energy potential would be significantly (50–75%) lower due to losses [16], [68] and [69]. Second-generation technologies for the production of liquid biofuels would be capable of using a considerably larger fraction of the plant materials available for bioenergy production, but would also involve conversion losses. A recent assessment recommends to favour direct use of solid plant materials over conversion to liquids, primarily based on comparisons of the GHG balances of different technologies [16].

Our assumption to base our estimates of bioenergy potentials on current (grazing areas) or potential (cropland) NPP (section 2.5; other recent studies [68] and [69] used similar assumptions) is also a simplification that might result in over- or underestimation of the potential. At present, the actual aboveground NPP on cropland and grazing areas is considerably lower than the potential NPP of these areas in the global average [38]. However, it would probably be possible to raise the NPP of bioenergy crops above the potential NPP of the areas on which they are planted through irrigation, fertilization, and other agricultural technologies, at least in many regions. While this might increase the amount of plant material produced, it would probably also result in a deterioration of the energy return on investment (EROI) and could lead to reduced, if not negative net energy gains [89] and [90]. Economic (agricultural investments) as well as biophysical (soil degradation, water availability) factors might also limit yield gains [64], [85] and [91]. We conclude that our bioenergy potential estimates could be regarded as a realistic to conservative: while we assume increases over current productivity levels, we do not assume massive intensification.

Comparison with Other Assessments of Bioenergy Potentials

Our bioenergy potential calculations do not include bioenergy potentials from forests. In the year 2000, the amount of wood fuels harvested in forests had an energy value of approximately 22 EJ [50]. The IEA reports that the total amount of "primary solid biomass" used for energy production globally was 39.4 EJ [92] and [93]. No comprehensive data exist to identify how much of the bioenergy currently used by humans comes from forests, from wastes in production processes, and from cropland and grazing areas. The potentials identified in this study include the unknown amount of bioenergy produced currently on cropland and grazing areas. The potential to produce bioenergy from forests was recently quantified to range from zero to 71 EJ y^{-1} in the year 2050: the global technical potential for forest bioenergy in 2050 was found to be 64 EJ y^{-1}, the economic potential 15 EJ y^{-1}, the ecological potential 8 EJ y^{-1} and the combined economic-ecological 0 EJ y^{-1} [18].

Table 7 compares the results of this study on global bioenergy potentials with current global biomass flows, with the current level of energy use, and with other studies on global bioenergy potentials. It shows that humans currently harvest and use a total amount of biomass with an energy value (GCV) of about 225 EJ y^{-1}, and that the total amount of biomass harvested, destroyed or burned due to human activities currently is around 310 EJ y^{-1}. This is a considerable fraction of the current aboveground NPP which is approximately 1241 EJ y^{-1}. These figures indicate that the primary bioenergy potential identified in this study (64-161 EJ y^{-1}) is considerable when compared to the current levels of human harvest and use of biomass or to current aboveground NPP.

Table 7: Current and projected future level of global biomass and energy use and global terrestrial net primary production: A compilation of estimates

	Energy flow [EJ y^{-1}]	Year	Sources
1. Current global NPP and its use by humans (gross calorific value)			
Total NPP of plants on earth's land	2191	2000	[38]
Aboveground NPP of plants on earth's land	1241	2000	[38]
Human harvest of NPP including by-flows, total	346	2000	[38] and [49]
Human harvest of NPP including by-flows, aboveground	310	2000	[38] and [49]
NPP harvested and actually used by humans	225	2000	[38] and [49]
2. Global human technical energy use (physical energy content)			
Fossil fuels (coal, oil, natural gas), gross calorific value	453	2008	[101]^a
Nuclear heat (assumed efficiency of nuclear plants 33%)	30	2008	[101]
Hydropower (assumed efficiency 100%)	11	2008	[101]
Wind, solar and tidal energy (100% efficiency)	1	2006	[102]
Geothermal (10% efficiency for electricity, 50% for heat)	2	2006	[102]
Biomass, including biogenic wastes, gross calorific value	54	2006	[102]^b
Total (physical energy content, gross calorific value)	551	2006–2008	[101] and [102]
3. Estimates of global bioenergy potentials or scenarios 2050 (calorific value not standardized)			
Bioenergy crops and residues, excluding forestry, this study	64–161	2050	
Mid-term potential according to the World Energy Assessment	94–280	2050	[10]
Review of mid-term potentials according to Berndes et al.	35–450	2050	[27]
Mid-term potential according to Fischer/ Schrattenholzer	370–450	2050	[103]
Potential according to Hoogwijk	33–1135	2050	[104]
IPCC-SRES scenarios mid-term	52–193	2050	[72]
Bioenergy potential on abandoned farmland	27–41	2050	[69] and [68]

Bioenergy potentials in forests	0–71	2050	[18]
Surplus agricultural land (not needed for food & feed)	215–1272	2050	[19]
Bioenergy crops (second generation)	34–120	2050	[16]

[a]BP reports energy data in tons of oil equivalent (toe) net calorific value (NCV). We assumed that 1 toe = 41.868 GJ (NCV). Conversion from NCV to gross calorific value (GCV) was based on the following multipliers (GCV/NCV): coal 1.1, oil 1.06, natural gas 1.11 [105].

[b]The IEA reports biomass as NCV; we converted this to GCV using a multiplier of 1.1.

The second part of Table 7 reveals, however, that the potential contribution of bioenergy from cropland and grazing areas is only a fraction of current fossil-fuel use. As shown in the lower part of Table 7, our estimate is considerably lower than the bioenergy potentials identified in many previous studies. We note that our estimate of primary bioenergy potential on cropland and grazing land is very similar to that of the WBGU[16], despite the fact that the methodology used by the WBGU was completely different from the one used here, but significantly lower than that found in other studies that did not consider links between food, feed and bioenergy.

CONCLUSIONS AND RECOMMENDATIONS

We conclude that the bioenergy potential on agricultural land in 2050 is highly sensitive to climate change as well as to changes in yields and diets. More research is required to better understand feedbacks between management, changes in precipitation, temperature, and the magnitude of the CO_2 fertilization effect under field conditions, all of which have a strong effect on the bioenergy potential. Our results suggest that the magnitude of global bioenergy potentials in the year 2050 is strongly affected by the need to produce feed for livestock, and that the careful consideration of biomass flows in the food system, in particular in the livestock system, is highly important in deriving realistic potentials for future bioenergy supply. Our results suggest that the bioenergy potential on agricultural areas in 2050 might be in the order of magnitude of 100 EJ y^{-1} based on current diet trajectories and a 'food first' approach; if 'poorer' diets are chosen, the potential may rise by up to 60%. A considerable fraction of this potential comes from agricultural residues, suggesting that in-depth assessments of options to combine bioenergy production and soil fertility management (e.g., energy production through biogas production that maintains a large proportion of the nutrients and parts of the carbon) should be undertaken. An integrated optimization of food and energy production based on a "cascade utilization" of biomass is an important option to produce and use bioenergy sustainably [16], [94] and [95]. Bioenergy potentials on grazing land, as calculated in this study, are substantial, but realizing them might entail massive investments in agricultural technology, such as irrigation infrastructure, and might be associated with vast social and ecological effects, such as a further pressure on populations that practice low-input agriculture. Realizing this potential might also trigger land-use change such as deforestation in far distant regions if not combined with robust measures to prevent such effects [17], [96] and [97]. At least at present, growth in agricultural production is a strong driver of deforestation [98].

ACKNOWLEDGEMENTS

This research was funded by the Austrian Science Fund (FWF) within the projects P20812-G11 and P21012-G11 as well as by Friends of the Earth, UK and the Compassion in World Farming Trust, UK. It contributes to the Global Land Project (http://www.globallandproject. org), to long-term socio-ecological research (LTSER) initiatives within LTER Europe (http://www.lter-europe.ceh.ac.uk/) and to the Global Energy Assessment of IIASA (http://www.iiasa.ac.at/Research/ENE/ GEA/index_gea.html). We thank an anonymous reviewer for valuable comments.

REFERENCES

1.   Millennium Ecosystem Assessment. Ecosystems and human well-being - our human Planet. Washington, D.C.: Island Press; 2005.

2. GLP. Global Land Project. Science plan and implementation strategy. Stockholm: IGBP Secretariat; 2005.

3. Turner BL, Lambin EF, Reenberg A. The emergence of land change science for global environmental change and sustainability. Proc Natl Acad Sci 2007;104:20666e71.

4. Foley JA, DeFries R, Asner GP, Barford C, Bonan G, Carpenter SR, et al. Global consequences of land use. Science 2005;309:570e4.

5. IPCC. Climate Change 2007-Impacts, adaptation andVulnerability. Contribution of working group II to the fourth assessment report of the IPCC. Cambridge: Cambridge University Press; 2007.

6. Long SP, Anisworth EA, Leakey ADB, No¨ sberger J, Ort DR. Food for thought: lower-than-exepted crop yield stimulation with rising CO2 concentrations. Science 2006; 312:918e21.

7. Tubiello FN, Soussana JF, Howden SM. Climate change and food security special feature: crop and pasture response to climate change. Proc Natl Acad Sci 2007;104:19686e90.

8. Fischer G, Shah M, Tubiello FN, Velhuizen Hv. Socioeconomic and climate change impacts on agriculture: an integrated assessment, 1990e2080. Phil Trans Royal Soc B Biol Sci 2005;360:2067e83.

9. Johansson TB, Kelly H, Reddy AKN, Williams RH. Renewable energy, sources for fuels and electricity. London, Washington D.C: Earthscan, Island Press; 1993.

10. Turkenburg WC. Renewable energy technology. In: Goldemberg J, editor. World energy assessment: energy and the challenge of sustainability. , New York: United Nations Development Programme (UNDP), United Nations Department of Economic and Social Affairs, World Energy Council (WEC); 2000. p. 219e72.

11. Jaccard M. Sustainable fossil fuels: the unusual suspect in the quest for clean and enduring energy. Cambridge: Cambridge University Press; 2005.

12. IPCC. Climate Change 2007-Mitigation. Contribution of working group III to the fourth Assessment Report of the IPCC. Cambridge: Cambridge University Press; 2007.

13. Rhodes JS, Keith DW. Biomass with capture: negative emissions within social and environmental constraints: an editorial comment. Clim Change 2008;87:321e8.

14. Lenton TM, Held H, Kriegler E, Hall JW, Lucht W, Rahmstorf S, et al. Tipping elements in the earth's climate system. Proc Natl Acad Sci 2008;105:1786e93.

15. Kriegler E, Hall JW, Held H, Dawson R, Schellnhuber H-J. Imprecise probability assessment of tipping points in the climate system. Proc Natl Acad Sci 2009;106:5041e6.

16. WBGU. Future bioenergy and sustainable land use. London: Earthscan; 2009.

17. Searchinger T, Heimlich R, Houghton RA, Dong F, Elobeid A, Fabiosa J, et al. Use of U.S. Croplands for biofuels increases greenhouse gases through emissions from land-use change. Science 2008;319:1238e40.

18. Smeets EMW, Faaij APC. Bioenergy potentials from forestry in 2050. Clim Change 2007;81:353e90.

19. Smeets EMW, Faaij APC, Lewandowski IM, Turkenburg WC. A bottom-up assessment and review of global bio-energy potentials to 2050. Progr Energy Combust Sci 2007;33:56e106.

20. UNEP. Assessing biofuels - Towards Sustainable production and use of resources. Paris: United Nations Environment Programme (UNEP), International Panel for Sustainable Resource Management, http://www.unep.fr; 2009.

21. Lapola DM, Schaldach R, Alcamo J, Bondeau A, Koch Koelking C, Priess JA. Indirect land-use changes can overcome carbon savings from biofuels in Brazil. Proc Natl Acad Sci 2010;107:3388e93.

22. World Bank. World development report 2008: agriculture for development. Washington, D.C.: The World Bank; 2008.

23. Oecd. Biofuel Support Policies, An economic assessment. Paris: Organization for Economic Co-Operation and Development (OECD); 2008.

24. Fischer G, Hizsnyik E, Prieler S, Shah M, Velthuizen Hv. Biofuels and food Security. Vienna, Laxenburg, Austria: OPEC Fund for International Development (OFID), International Institute for Applied Systems Analysis (IIASA); 2009.

25. EEA. How much bioenergy can Europe produce without harming the environment? Copenhagen: European Environment Agency; 2006.

26. Gerbens-Leenes PW, Hoekstra AY, Meer Thvd. The water footprint of energy from biomass: a quantitative assessment and consequences of an increasing share of bio-energy in energy supply. Ecol Econ 2009;68:1052e60.

27. Berndes G, Hoogwijk M, van den Broek R. The contribution of biomass in the future global energy supply: a review of 17 studies. Biomass Bioenerg 2003;25:1e28.

28. Hoogwijk M, Faaij A, Eickhout B, de Vries B, Turkenburg W. Potential of biomass energy out to 2100, for four IPCC SRES land-use scenarios. Biomass Bioenerg 2005;29:225e57.

29. Sims REH, Hastings A, Schlamadinger B, Taylor G, Smith P. Energy crops: current status and future prospects. Glob Change Biol 2007;12:2054e76.

30. Ayres RU, Simonis UE, editors. Industrial metabolism: Restructuring for Sustainable development. Tokyo, New York, Paris: United Nations University Press; 1994.

31. Ayres RU, Ayres LW. Accounting for resources, 1, EconomyWide applications of mass-balance principles to materials and Waste. Cheltenham, UK: Edward Elgar; 1998.

32. Fischer-Kowalski M. Society's metabolism. The Intellectual History of material flow analysis, Part I: 1860e1970. J Ind Ecol 1998;2:61e78.

33. Fischer-Kowalski M, Hüttler W. Society's metabolism. The Intellectual History of material flow analysis, Part II: 1970e1998. J Ind Ecol 1998;2:107e37.

34. Martinez-Alier J. Social metabolism, ecological distribution conflicts, and Languages of Valuation. Capitalism Nat Socialism 2009;20:58e87.

35. Vitousek PM, Ehrlich PR, Ehrlich AH, Matson PA. Human appropriation of the products of photosynthesis. BioSci 1986;36:363e73.

36. Haberl H. Human appropriation of net primary production as an environmental indicator: implications for sustainable development. Ambio 1997;26:143e6.

37. Wright DH. Human impacts on the energy flow through natural ecosystems, and implications for species endangerment. Ambio 1990;19:189e94.

38. Haberl H, Erb K-H, Krausmann F, Gaube V, Bondeau A, Plutzar C, et al. Quantifying and mapping the human appropriation of net primary production in earth's terrestrial ecosystems. Proc Natl Acad Sci 2007;104:12942e7.

39. FAO. World agriculture: towards 2030/2050-Interim report. Prospects for food, nutrition, agriculture and major commodity groups. Rome: Food and Agricultural Organization (FAO); 2006.

40. Bondeau A, Smith PC, Zaehle S, Schaphoff S, Lucht W, Cramer W, et al. Modelling the role of agriculture for the 20th century global terrestrial carbon balance. Glob Change Biol 2007;13:679e706.

41. UNSD. Composition of macro geographical (continental) regions, geographical sub-regions, and selected economic and other groupings. New York: United Nations Statistical Division, http:// unstats.un.org/unsd/methods/m49/ m49regin.htm; 2006.

42. Krausmann F, Haberl H, Erb K-H, Wiesinger M, Gaube V, Gingrich S. What determines geographical patterns of the global human appropriation of net primary production? J Land Use Sci 2009;4:15e34.

43. Erb KH, Krausmann F, Lucht W, Haberl H. Embodied HANPP: Mapping the spatial disconnect between global biomass production and consumption. Ecol Econ 2009;69(2): 328e34.

44. Souci SW, Fachmann W, Kraut H. Food composition and nutrition tables. Boca Raton: CRC; 2000.

45. Purdue University Center for New Crops and Plant Products. Crop index database. Purdue University, http://www.hort. purdue.edu/ newcrop/Indices/index_ab.html; 2006.

46. Lo"hr L. Faustzahlen fu" r den Landwirt. Graz, Stuttgart: Leopold Stocker Verlag; 1990.

47. Watt BK, Merrill AL. Handbook of the nutritional contents of foods. New York: Dover Publications; 1975.

48. Erb KH, Gaube V, Krausmann F, Plutzar C, Bondeau A, Haberl H. A comprehensive global 5min resolution land-use dataset for the year 2000 consistent with national census data. J Land Use Sci 2007;2:191e224.

49. Krausmann F, Erb K-H, Gingrich S, Lauk C, Haberl H. Global patterns of socioeconomic biomass flows in the year 2000:

a comprehensive assessment of supply, consumption and constraints. Ecol Econ 2008;65:471e87.

50. FAO. FAOSTAT 2004, FAO statistical databases: agriculture, Fisheries, forestry, Nutrition. Rome: FAO; 2004.

51. UN. Forest resources of europe, CIS, North America, Australia, Japan and New Zealand (industrialized temperate/boreal countries). UN-ECE/FAO contribution to the global forest resources assessment 2000. New York, Geneva: United Nations Publications; 2000.

52. Lauk C, Erb K- H. Biomass consumed in anthropogenic vegetation fires: global patterns and processes. Ecol Econ 2009;69:301e9.

53. Sanderson E, Jaiteh M, Levy M, Redford K, Wannebo A, Woolmer G. The human footprint and the last of the wild. BioScience 2002;52:891e904.

54. Erb K-H, Haberl H, Krausmann F, Lauk C, Plutzar C, Steinberger JK, et al. Feeding and fuelling the world sustainably, fairly and humanely, a scoping study. Vienna: Institute of Social Ecology; 2009. Working Paper No. 116.

55. Wirsenius S. Human use of land and Organic materials. Modeling the Turnover of biomass in the global food system. Go¨teborg: Chalmers University; 2000.

56. Wirsenius S. The biomass metabolism of the food system. A model-based Survey of the global and regional Turnover of food biomass. J Ind Ecol 2003;7:47e80.

57. UN. World population prospects: the 2006 revision. New York: United Nations, Dept. Econ. Soc. Affairs, Popul. Div; 2007.

58. FAO. FAOSTAT 2005, FAO statistical databases: agriculture, Fisheries, forestry, Nutrition. Rome: FAO; 2005.

59. Alexandratos N. World agriculture: towards 2010. An FAO Study. Chichester, New York: FAO/John Wiley & Sons; 1995.

60. Bruinsma J. World agriculture: towards 2015/2030. An FAO perspective. London: Earthscan; 2003.

61. Fischer G, Velthuizen Hv, Nachtergaele FO. Global agroecological zones assessment: methodology and results. IIASA interim report IR-00e064. Laxenburg: IIASA; 2000.

62. Rosegrant MW, Paisner MS, Meijer S, Witcover J. Global food projections to 2020: Emerging trends and alternative futures. Intl Food Policy Res Inst; 2001.

63. Tilman D, Fargione J, Wolff B, D'Antonio C, Dobson A, Howarth R, et al. Forecasting agriculturally driven global environmental change. Science 2001;292:281e4.

64. IAASTD. Agriculture at a Crossroads. International assessment of agricultural knowledge, science and technology for development (IAASTD), global report. Washington, D.C.: Island Press; 2009.

65. IIASA, Fao. Global agro-ecological Zones 2000. Rome, Italy: International Institute for Applied Systems Analysis (IIASA) and Food and Agriculture Organization (FAO); 2000.

66. Ramankutty N, Foley JA, Norman J, McSweeney K. The global distribution of cultivable lands: current patterns and sensitivity to possible climate change. Glob Ecol Biogeogr 2002;11:377e92.

67. Haberl H, Erb K- H. Assessment of sustainable land use in producing biomass. In: Dewulf J, Langenhove HV, editors. Renewables-Based technology: sustainability assessment. Chichester: Wiley; 2006. p. 175e92.

68. Field CB, Campbell JE, Lobell DB. Biomass energy: the scale of the potential resource. Trends Ecol Evol 2008;23:65e72.

69. Campbell JE, Lobell DB, Genova RC, Field CB. The global potential of bioenergy on abandoned agriculture lands. Environ Sci Technol 2008;42:5791e5.

70. Lal R. World crop residues production and implications of its use as a biofuel. Environ Int 2005;31:575e84.

71. Fader M, Rost S, Mu¨ller C, Bondeau A, Gerten D. Virtual water content of temperate cereals and maize: present and potential future patterns. J Hydrol 2010;384:218e31.

72. Nakicenovic N, Swart R. In: Special report on emission scenarios. Cambridge: Cambridge University Press; 2000.

73. Collins WD, Bitz CM, Blackmon ML, Bonan GB, Bretherton CS, Carton JA, et al. The Community climate system model version 3 (CCSM3). J Clim 2006;19:2122e43.

74. Jungclaus JH, Keenlyside N, Botzet M, Haak H, Luo JJ, Latif M, et al. Ocean circulation and tropical variability in the coupled model ECHAM5/MPI-OM. J Clim 2006;19:3952e72.

75. Min SK, Legutke S, Hense A, Kwon WT. Internal variability in a 1000-yr control simulation with the coupled climate model ECHO-G - I. Near-surface temperature, precipitation and mean sea level pressure. Tellus A 2005;57: 605e21.

76. Delworth TL, Broccoli AJ, Rosati A, Stouffer RJ, Balaji V, Beesley JA, et al. GFDL's CM2 global coupled climate models. Part I: Formulation and simulation characteristics. J Clim 2006;19:643e74.

77. Cox PM, Betts RA, Bunton CB, Essery RLH, Rowntree PR, Smith J. The impact of new land surface physics on the GCM simulation of climate and climate sensitivity. Clim Dynam 1999;15:183e203.

78. New MG, Hulme M, Jones PD. Representing twentiethcentury space-time climate variability. Part II: development of 1901-1996 monthly grids of terrestrial surface climate. J Clim 2000;13:2217e38.

79. Ö sterle H, Gerstengarbe FW. Homogenisierung und Aktualisierung des Klimadatensatzes des Climate Research Unit der Universita¨t of East Anglia. Norwich: University of East Anglia; 2003.

80. Young A. Is there Really Spare land? A Critique of estimates of available cultivable land in developing countries. Environ Devel Sustain 1999;1:3e18.

81. Showers KB. A History of African soil: Perceptions, use and Abuse. In: McNeill JR, Winiwarter V, editors. Soils and Societies. Perspectives from environmental history. Cambridge: The White Horse Press; 2006. p. 118e76.

82. Cassman KG. Ecological intensification of cereal production systems: yield potential, soil quality, and precision agriculture. Proc Natl Acad Sci 1999;96:5952e9.

83. Peng S, Laza RC, Visperas RM, Sanico AL, Cassman KG, Khush GS. Grain yield of rice cultivars and lines developed in the Philippines since 1966. Crop Sci 2000;40:307e14.

84. Tilman D, Cassman KG, Matson PA, Naylor R, Polasky S. Agricultural sustainability and intensive production practices. Nature 2002;418:671e7.

85. Kahn BM, Zaks D, Fulton M, Dominik M, Soong E, Baker J, et al. Investing in agriculture: far-Reaching challenge, significant

opportunity, an asset management perspective. Frankfurt: DB Climate Change Advisors, Deutsche Bank Group, http://www.dbcca.com/research; 2009.

86. Tubiello FN, Ewert F. Simulating the effects of elevated CO2 on crops: approaches and applications for climate change. Europ J Agron 2002;18:57e74.

87. Taub DR, Miller B, Allen H. Effects of elevated CO2 on the protein concentration of food crops: a meta-analysis. Glob Change Biol 2008;14:565e75.

88. Zavala JA, Casteels CL, DeLucia EH, Berenbaum MR. Anthropogenic increase in carbon dioxide compromises plant defense against invasive insects. Proc Natl Acad Sci 2008;105:5129e33.

89. Giampietro M, Mayumi K. The Biofuel Delusion. The fallacy of large-scale agro-biofuel production. London: Earthscan; 2009.

90. Pimentel D, editor. Biofuels, Solar and Wind as Renewable energy systems. Benefits and risks. New York: Springer; 2008.

91. Gerten D, Rost S, Bloh W, Lucht W. Causes of change in 20th century global river discharge. Geophys Res Lett 2008;35. doi:10.1029/2008GL035258.

92. IEA. Energy statistics of Non-OECD countries, 2004e2005. CD-ROM. Paris: International Energy Agency (IEA), Organisation of Economic Co-Operation and Development (OECD); 2007.

93. IEA. Energy statistics of OECD countries, 2004e2005. CDROM. Paris: International Energy Agency (IEA), Organisation of Economic Co-Operation and Development (OECD); 2007.

94. Haberl H, Geissler S. Cascade Utilisation of Biomass: how to cope with ecological limits to biomass use. Ecol Eng 2000; 16(Suppl.):S111e21.

95. Haberl H, Erb KH, Krausmann F, Adensam H, Schulz NB. Land-Use change and socioeconomic metabolism in Austria. Part II: land-use scenarios for 2020. Land Use Pol 2003;20:21e39.

96. Fargione J, Hill J, Tilman D, Polasky S, Hawthorne P. Land Clearing and the biofuel carbon Debt. Science 2008;319: 1235e8.

97. Koh LP, Ghazoul J. Biofuels, biodiversity, and people: understanding the conflicts and finding opportunities. Biol Conserv 2008;141:2450e60.

98. Geist HJ, Lambin EF. What Drives Tropical Deforestation? A meta-analysis of proximate and underlying causes of deforestation based on subnational case study evidence. Louvain-la-Neuve; 2001. LUCC Report Series No. 4.

99. FAO. Statistical databases: agriculture, Fisheries, forestry, Nutrition. . Rome: Food and Agriculture Organization of the United Nations (FAO); 2006.

100. UN Statistics Division. National accounts main aggregates. New York: United Nations; 2006.

101. BP. Statistical review of world energy 2009. British Petroleum (BP), http://www.bp.com/statisticalreview; 2009. London.

102. IEA. Renewables information 2008. Paris: International Energy Agency (IEA), Organization for Economic CoOperation and Development (OECD); 2008.

103. Fischer G, Schrattenholzer L. Global bioenergy potentials through 2050. Biomass Bioenerg 2001;20:151e9.

104. Hoogwijk M, Faaij A, Rvd Broek, Berndes G, Gielen D, TurkenburgW. Exploration of the ranges of the global potential of biomass for energy. Biomass Bioenerg 2003;25:119e33.

105. Haberl H, Weisz H, Amann C, Bondeau A, Eisenmenger N, Erb K-H, et al. The energetic metabolism of the EU-15 and the USA. Decadal energy input time-series with an emphasis on biomass. J Ind Ecol 2006;10:151e71.

Chapter 2

Development of an Estimation Model for the Evaluation of the Energy Requirement of Dilute Acid Pretreatments of Biomass

Oluwakemi A.T. Mafe[a], Scott M. Davies[b], John Hancock[b], and Chenyu Du[a]

[a]School of Biosciences, Sutton Bonington Campus, University of Nottingham, Loughborough LE12 5RD, United Kingdom

[b]Briggs of Burton, Briggs House, Derby Street, Burton on Trent, Staffordshire DE14 2LH, United Kingdom

ABSTRACT

This study aims to develop a mathematical model to evaluate the energy required by pretreatment processes used in the production of second generation ethanol. A dilute acid pretreatment process reported by National Renewable Energy Laboratory (NREL) was selected as an example for the model's development. The energy demand of the pretreatment process was evaluated by considering the change of internal energy of the substances, the reaction energy, the heat lost and the work done to/by the system based on a number of simplifying assumptions. Sensitivity analyses were performed on the solid loading rate, temperature, acid concentration and water evaporation rate. The results from the sensitivity analyses established that the solids loading rate had the most significant impact on the energy demand. The model was then verified with data from the NREL benchmark process. Application of this model on other dilute acid pretreatment processes reported in the literature illustrated that although similar sugar yields were reported by several studies, the energy required by the different pretreatments varied significantly.

INTRODUCTION

The world population is expected to reach 9.6 billion by 2050 and will demand a large amount of energy to allow these people to fulfil their daily lives. In approximately 20% of the world's population (predominantly in the under-developed nations), continuous and reliable supply of energy is not easily accessible [1]. These under-developed nations seek to improve their standard of living by tapping into the existing energy resources. This in combination with the ever-increasing population puts a tremendous strain on the finite fossil fuel resources.

Lignocellulosic materials including for example agricultural residues such as bagasse [2], corn stover [3] and wheat straw [4]; forest residues [5]; energy crops [6] and waste paper [7] provide a renewable and potentially inexpensive source of raw material for the production of liquid fuels such as ethanol. However, to effectively gain access to the sugars in the lignocellulose structure, a pretreatment process is required to weaken the naturally recalcitrant structure of lignocellulosic materials [8], [9], [10], [11], [12] and [13]. Pretreatment processes are currently essential for the conversion of lignocellulosic materials into ethanol using biocatalyst such as enzymes and fermentative microorganisms. Pretreatments are recognised as a large contributor to the cost of cellulosic ethanol production. In some processes, pretreatment is responsible for up to 14% of the total fixed capital

[14] while Yang and Wyman proposed that pretreatment accounted for 20% of the total cost of production [8].

Several studies have been carried out on the economic performance [15], [16] and [17], the life cycle assessment [18], and the minimum selling price of the ethanol [16] and [17] of the various pretreatments but very few have considered the evaluation of the energy consumption in the pretreatment process. Conde-Mejia et al. [19], using ASPEN software compared the energy consumption and associated cost of several pretreatment processes with and without direct recycle streams implemented into the design and concluded that steam explosion and dilute sulphuric acid pretreatments were the most energy efficient pretreatments in terms of the energy cost (dollars per tonne of dry biomass). Kumar and Murthy [20] used the simulation software SuperPro Designer to explore the energy balance of the production of cellulosic ethanol and also found steam explosion to be the most energy efficient pretreatment; its ethanol yield however was the lowest of the pretreatment processes examined. These studies however contain insufficient information on the key influences of the rate of energy consumption during the pretreatment process, which is a key factor in evaluating the economic performance of the process. These studies focussed more on the types of processing technique used, processing equipment used, operating conditions and the price of ethanol (15–19).

This paper therefore aims at presenting an effective energy estimation model developed from a fundamental level to clearly identify the energy consumption factors of a pretreatment process. In this case, the dilute acid pretreatment process was used as an example.

MODEL DEVELOPMENT AND RESULTS

Process Description of the Base Case

The pretreatment design basis for the energy estimation model was taken from the National Renewable Energy Laboratory (NREL) dilute acid process [21]. The process was built upon the evaluation of previous models designed by NREL. It should be noted that the original operating conditions of 190 °C for 2 min with a sulphuric acid mass fraction of 1.1% was deemed too harsh and resulted in a significant number of degradation products. NREL subsequently revised the process to reduce the pretreatment severity without generating high amounts of the degradation products. This included a two-stage pretreatment process. Stage 1 uses a sulphuric acid content per dry g of biomass of 18 mg at 158 °C and 557 kPa (5.5 atm) for a period of 5 min with a solids fraction of 30% (defined as the amount of dry biomass divided by the total mass of the biomass and liquid added). Under these conditions a considerable amount of oligomers is formed from the glucan, xylan, arabinan, mannan and galactan from the plant cell wall hemicelluloses. The second stage of the pretreatment hydrolysis is operated at 130 °C for 20 min–30 min with the addition of a further 4.1 mg of sulphuric acid. This hydrolyses the oligomers released in the first stage into their respective monomers (glucose, xylose, arabinose, mannose and galactose).

During this process, there are several chemical reactions that occur [21]. Table 1 lists the chemical reactions that were considered for the model in this paper. Arabinan, mannan, and galactan were assumed to have the same reactions and conversion pathways as xylan. The reactions involving the formation of inhibitory compounds from mannan and galactan were assumed to form 5-hydroxymethylfurfural (HMF) as in the case of glucan.

Table 1: Pretreatment hydrolysis reactions of the NREL dilute acid pretreatment process with the compositions of the reactants, the reactions' assumed conversions and the calculated heats of formation values

No.	Reactions[a]	Composition[b]	Conversions[a]	Heats of Formation/ cal mol^{-1c}
1	(Glucan)$_n$ + nH$_2$O → nGlucose	31.9%	9.90%	1004
2	(Glucan)$_n$ + nH$_2$O → nGlucose Oligomer	31.9%	0.30%	68 232
3	(Glucan)$_n$ → nHMF + 2nH$_2$O	31.9%	0.30%	−2941
4	Sucrose → HMF + Glucose + 2H$_2$O	3.6%	100%	−55,669
5	(Xylan)$_n$ + nH$_2$O → nXylose	18.9%	90.0%	892

6	$(Xylan)_n + mH_2O \rightarrow mXylose$ Oligomer	18.9%	2.40%	0
7	$(Xylan)_n \rightarrow nFurfural + 2nH_2O$	18.9%	5.00%	−2102
8	$Acetate \rightarrow Acetic\ Acid$	2.2%	100%	26
9	$(Lignin)_n \rightarrow nSoluble\ Lignin$	13.3%	5.00%	0
10	$(Arabinan)_n + nH_2O \rightarrow$ nArabinose	2.8%	90.0%	892
11	$(Arabinan)_n + mH_2O \rightarrow$ mArabinose Oligomer	2.8%	2.40%	0
12	$(Arabinan)_n \rightarrow nFurfural + 2nH_2O$	2.8%	5.00%	−2102
13	$(Mannan)_n + nH_2O \rightarrow$ Mannose	0.3%	90.0%	1004
14	$(Mannan)_n + mH_2O \rightarrow$ mMannose Oligomer	0.3%	2.40%	0
15	$(Mannan)_n \rightarrow nHMF + 2nH_2O$	0.3%	5.00%	−2941
16	$(Galactan)_n + nH_2O \rightarrow$ nGalactose	1.5%	90.0%	1004
17	$(Galactan)_n + mH_2O \rightarrow$ mGalactose Oligomer	1.5%	2.40%	0
18	$(Galactan)_n \rightarrow nHMF + 2nH_2O$	1.5%	5.00%	−2941

Process Model Development and Calculations

In this study, an energy model was developed to simulate the NREL pretreatment process. It contained five stages according to temperature changes as shown in Fig. 1 and Table 2. These five stages are outlined as follows; Stage 1, the supply of steam to the pretreatment reactor in order to attain the desired reaction temperature of 158 °C; Stage 2, the supply of energy to maintain this temperature for the specified residence time of 5 min during which the oligomers are formed; Stage 3, the release of energy in order to cool the contents of the outlet stream down to 130 °C, the temperature required for the conversion of the oligomers; Stage 4, the supply of energy to maintain the reactor at 130 °C and allow the conversion of oligomers to monomers and Stage 5, the release of energy in order to cool the products down to 97 °C; the temperature suited for the next stage of the ethanol production process.

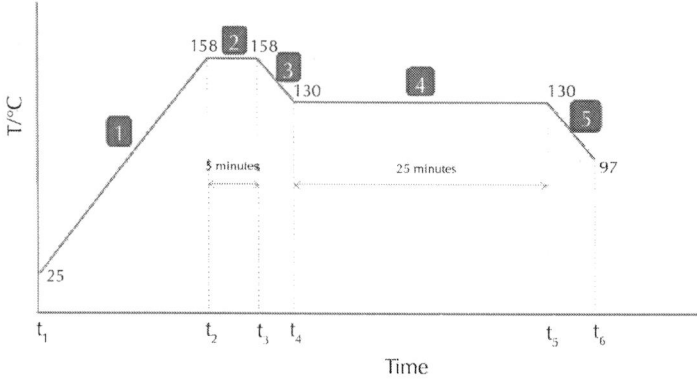

Figure 1: Temperature profile of the NREL dilute acid pretreatment process.

Table 2: Operating conditions of the pretreatment process

	Stage 1	Stage 2	Stage 3	Stage 4	Stage 5
T/°C	25 up to 158	158	158 down to 130	130	130 down to 97
Residence time/min	unknown	5	Unknown	25^a	unknown
p/kPa	–	557	–	–	101
Acid loading/ mg g^{-1b}	none	18	None	4.1	none

The energy estimation model created is an intricate version of the net energy balance Equation for a closed system;

$$\Delta E = Q - W \tag{1}$$

Where E is the change of the internal energy of the system, Q is the heat added to the system and W is the work done by the system. Equation (1) can be developed to incorporate the 5 stages outlined earlier as:

$$E = QStage_1 + QStage_2 + QStage_3 + QStage_4 + QStage_5 - WT \tag{2}$$

Where W_T represents the total work done over the whole system.

Stage 1 is otherwise known as the heating stage while both Stage 2 and Stage 4 represent the period in which the hydrolysis reactions occur. In addition, Stages 3 and 5 are the cooling stages. It is obvious that energy could be recovered from Stages 3 and 5. As the paper aims to estimate the theoretical energy requirement for the pretreatment process, and Stages 3 and 5 do not require an input of energy but rather a release, these two-stages were not modelled. Equation (2) is therefore further simplified to Equation (3):

$$E = Q_{heating} + Q_{reaction} - WT \qquad (3)$$

There is no work done by this pretreatment system and the agitation energy input into the system was ignored for simplicity purposes. The WT term was not considered in the following estimation.

It is inevitable that there will be some form of heat loss but in order to simplify the development of the model, this was set to zero. Its impact was however analysed during the model improvement stage.

The $Q_{heating}$ term in Equation (3) can be further broken down to Equation (4).

$$Q_{heating} = m \int_{T_0}^{T_H} C_{p\,w}\, dT + m_b \int_{T_0}^{T_H} C_{pb}\, dT + m_a\, \bar{C}_{pa}(T_H - T_0) \qquad (4)$$

C_p represents the specific heat capacity in kJ kg^{-1} °C^{-1}; m is mass in kg with the subscripts w, b and are presenting the water, biomass (corn stover) and sulphuric acid, respectively. T is the temperature in which T_H is the hydrolysis temperature in °C; and T_0 the environmental temperature set at 25 °C.

APPLICATION OF THE MODEL TO THE DILUTE ACID PRETREATMENT PROCESS

To model the stages considered, three key components were specified; corn stover, sulphuric acid and process water as denoted in Equation

(4). Some assumptions were also established to aid the development process and for the heating stage which include:

1. The model was created on a one-kilogram (1 kg) basis of dry corn stover at a solids mass fraction of 30%.

2. The environmental temperature of the pretreatment process was assumed to be 25 °C.

3. It was speculated that none of the chemical reactions outlined in Table 1 occurred during this stage.

4. It was also decided upon that the heating energy of the 3 components would be calculated individually and the sum used to simulate the total heating for this stage as depicted in Equation (4). The impact of mixing sulphuric acid with water on the heating energy was analysed in the model improvement section.

5. For the sulphuric acid heating energy, an average Cp value was used.

Heating Energy Calculation

The water heating energy was calculated using the first term in Equation (4) in which the specific heat capacity value used was obtained from the second order of the polynomial plot of varying water specific heat capacities against their corresponding temperature values as shown in Equation (5) (the specific heat capacity values were collected in the temperature range of 0.01 °C–200 °C from Engineering Toolbox [22])

$$C_{pw}=0.00001T^2-0.0013T+4.2085 \qquad (5)$$

As the energy required for heating the water from 25 °C to 158 °C was to be calculated, it was determined that at a pressure of 557 kPa (5.5 atm), water boils at 155 °C. As a result, the latent heat of vaporisation of water at 155 °C was adopted into the model. This revelation also introduced an expression for the heating of the water vapour from 155 °C to 158 °C. At this stage in the model however, the total percentage of water presumed to evaporate at 155 °C was 5% (this was later confirmed to be 4%). Consequently, the assumption introduces another expression for the heating of the remaining 95% of water from 155 °C to 158 °C. For the heating energy of the water vapour, its specific heat capacity was calculated using Equation (6), which was created on the manipulation of specific heat values obtained from Engineering Toolbox [23].

$$C_p = 0.0000008T^2 + 0.0002T + 1.8572 \qquad (6)$$

Factoring in all these aspects, the term for the water heating energy is therefore modified to:

$$m_w \int_{25}^{155} (0.00001T^2 - 0.0013T + 4.2085)dT + 0.05m_w H_L$$

$$+ (m_w - 0.05m_w) \int_{155}^{158} (0.00001T^2 - 0.0013T + 4.2085)dT$$

$$+ 0.05m_w \int_{155}^{158} (0.0000008T^2 + 1.8572)dT \qquad (7)$$

As a result, the total heating energy for 2.315 kg of water (mw) was found to be 1531 kJ kg^{-1}.

For the corn stover heating energy, Differential Scanning Calorimetry (DSC) experiments were carried out in triplicate to obtain a dataset of specific heat capacity values which were plotted against their corresponding temperature values to produce Equation (8). The corn stover analysed was obtained on August 12 2013 from a local corn farm in Loughborough (United Kingdom); all the parts of the corn plant minus the roots and the ears were collected. The corn stover was approximately 3 months old with an initial moisture content of 11.6% and average cross-section diameter of around 18 mm. The corn stover sample was stored in a dark room at room temperature and inert conditions until use. The sample was washed and oven-dried overnight before being pulverised in a Fritsch P5 Planetary Ball Mill to an almost homogenized sample. The milling process was carried out at 10 repetitions of 5 min milling and 5 min pausing after which the sample was sealed in an aluminium tray at ambient temperature until further use in the DSC (Perkin Elmer DSC7 – ZAAA0495). For the DSC experiments, a known mass of the sample was pelleted in aluminium pans and analysed to temperatures of 160 °C. Triplicate runs of the sample were performed from which heat flow curves were obtained and manipulated to give the specific heat capacity curves and hence equation for the locally grown corn stover:

$$Cp=0.00004T^2-0.0015T+0.9325 \qquad (8)$$

The corn stover heating energy was therefore calculated using Equation (9) which was developed from the substitution of Equation (8) into the second term in Equation (4). The result of which was 158 kJ kg^{-1} at $mb = 1$ kg.

$$m_b \int_{25}^{158} (0.00004T^2 - 0.0015T + 0.9325)dT \qquad (9)$$

Alternatively, the energy required for heating the sulphuric acid was calculated using the last term in Equation(4) with an average specific heat capacity value of 1.34 kJ kg^{-1} °C^{-1}[24] and a sulphuric acid mass (ma) of 0.018 kg. The outcome of this calculation was 3 kJ kg^{-1}; thus making the total heating energy of the dilute acid pretreatment process 1692 kJ kg^{-1}. Fig. 2 shows the distribution of this energy in relation to the three key components.

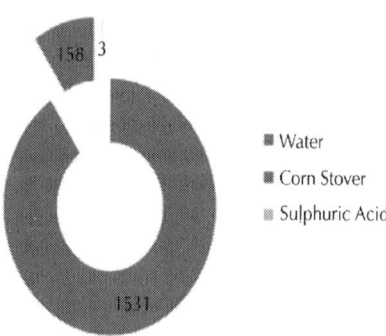

Figure 2: Illustration of the total pretreatment heating energy (kJ kg^{-1}).

From Fig. 2 it is clear that the main component responsible for the magnitude of the heating energy is water. This suggests that the solid loading rate is one of the determining factors that govern the amount of energy consumed in the heating stage of the pretreatment process. On the other hand, the amount of energy required to heat the acid up is negligible, indicating that in theory the sugar yield can be improved by the addition of acid without a notable increase in the energy consumed – presumably an advantage of the dilute acid pretreatment over other pretreatment processes like the hydrothermal.

Reaction Energy Calculation

For the reaction stage, three assumptions were made:

1. The reaction energy covered both stage 2 and stage 4 shown in Fig.
2. All the reactions specified in Table 1 occurred in this stage.

The reaction energy was therefore calculated using the information in Table 1 together with Equation (10):

$$Q_{reaction}=\sum(mb \times cr \times xr \times Hf/Mw) \qquad (10)$$

Where cr and xr are the composition of the reactant in the biomass and the conversion rate of the reactant to the product respectively and Mw the molecular weight of the reactant.

The equation was applied to the reactions and the overall net energy for the pretreatment process was found to be −17 kJ kg^{-1}. The negative energy value implies that the overall reaction process is exothermic and therefore in theory, results in a release of energy.

The total energy demand for the NREL dilute acid pretreatment process on a 1-kg basis according to the energy model is therefore only the heating energy of 1692 kJ kg^{-1}. In practice, a certain amount of energy could be recovered from the cooling stages. The net energy requirement will be lower than this value.

MODEL IMPROVEMENT

Some adaptations were made to look at the effect of calculating the heating energy of the sulphuric acid and water combined together on the total heating energy and to model the heat loss component that was initially ignored.

Sulphuric Acid-Water Heating Energy

In the model developed, an individual component heating method was implemented for ease of calculation. It was thought that the results from this method would not be far off from the results that would have been obtained if the heating energy for water and sulphuric acid were calculated together. In order to validate this assumption, the heating energy for the sulphuric acid and water combined together was calculated using a specific heat capacity graph for the sulphuric acid-water system obtained from a sulphuric acid bulletin [25].

A slightly modified version of Equations (4) and (11) was used in working out the heating energy involving the sulphuric acid-water system and is a modification of Equation (4), incorporating the combined mass of water and sulphuric acid and specific heat capacity of water and sulphuric acid:

$$Q_{heating} = m_b \int_{T_0}^{T_H} C_{pb} dT + m_{a-w} \cdot \overline{C}_{p_{a-w(T_H-T_0)}} \cdot (T_H - T_0) \tag{11}$$

Where subscript a–w represents the sulphuric acid-water system and the biomass heating term is kept constant.

In the heating stage of the pretreatment process (Stage 1), the acid loading rate per gram of dry corn stover is 18 mg, which is equivalent to a concentration of 0.7%; the boiling point of the sulphuric acid-water system was therefore assumed to be the same as pure water at 557 kPa (5.5 atm) (155 °C) due to the small concentration of acid and just as in the case of the individual heating energy of water there are four heating parts to the sulphuric acid-water heating energy; the heating of the system to its boiling point, the latent heat energy, the heating of the vapour to the reaction temperature and the heating of the remaining liquid to the reaction temperature. The expression for the sulphuric acid-water heating energy therefore becomes:

$$
\begin{aligned}
& m_{a-w} \cdot \overline{C}_{p_{a-w(155,25)}} \cdot (155 - 25) + 0.05 m_{a-w} \times H_L \\
& + (m_{a-w} - 0.05 m_{a-w}) \cdot \overline{C}_{p_{a-w(158,155)}} \cdot (158 - 155) \\
& + 0.05 m_{a-w} \cdot \overline{C}_{p_{a-w(158,155)}} \cdot (158 - 155)
\end{aligned}
\tag{12}
$$

The sulphuric acid-water system graph used was assumed to represent both liquid and vapour form of the sulphuric acid-water mixture at a mass fraction of 0.7%, simplifying the above heating energy equation to:

$$m_{a-w} \cdot \overline{C}_{p_{a-w(155,25)}} \cdot (155-25) + 0.05 m_{a-w} \times H_L + (m_{a-w} \cdot \overline{C}_{p_{a-w(158,155)}} \cdot (158-155) \quad (13)$$

The outcome of these calculations was 1541 kJ kg^{-1}, which is close to the 1534 kJ kg^{-1} calculated in the individual heating method; thereby justifying the assumption made about the method of calculation chosen in the model. It should be noted however that the same latent heat of vaporisation of water at 155 °C was used.

Heat Loss

To model the heat loss during the pretreatment process, three different methods were used. In the first approach, the amount of heat lost was evaluated from an experimental procedure using a Genesis Benchtop Autoclave by Rodwell Scientific Instruments to simulate the pretreatment process. The autoclave was run at 123 °C for 18 min. The entire experiment ran for approximately 115 min with the power readings taken with a plug-in mains power and energy monitor from Maplin Electronics (L61AQ) every 5 min for the heating and cooling stages which lasted 40 and 55 min respectively and every minute for the 'hydrolysis' stage. During the heating up period, the power readings remained relatively constant at 2.4 kW while during the 'hydrolysis' stage, the power readings fluctuated approximately every 6 s between high (2.4 kW) and low (0.013 kW) readings. Therefore, the power requirement for the maintenance was estimated to be 50% of that of the heating stage. Given that the residence time for the maintenance stage in the model is 5 min and assuming a heating up time of 15 min, the heat loss during the hydrolysis stage in this approach was calculated using Equation (14):

$$Q_{lost} = 0.5 \times \left(\frac{T_{maintenance}}{T_{heating}} \right) \times Q_{heating} \quad (14)$$

This resulted in a heat loss value of 282 kJ kg^{-1}.

In the second approach, the pretreatment reactor was assumed to be of an insulated cylindrical nature and as a result, the amount of heat lost during the pretreatment process was likened to that of a pipe where the heat loss is represented as:

$$Qlost = UA\ T \qquad (15)$$

As the heat transfer is considered to be through the wall of an insulated pipe, the overall heat transfer coefficient, U, is defined as:

$$U = 1/[(R_3/R_1h_{in}) + (R_2\ ln(R_2\ ln(R_2/R_1)/k_{pipe}$$
$$+ (R_3\ ln(R_3\ ln(R_3/R_2)/k_{insulation}) + 1/h_{out}] \qquad (16)$$

Generally, the first two terms of the denominator in Equation (16) are a lot smaller than the other two terms and so their influence on the result of the final calculation of overall heat transfer coefficient would be insignificant. The equation is therefore simplified to:

$$U = 1/[(R_3\ ln(R_3/R_2)/k_{insulation}) + 1/h_{out}] \qquad (17)$$

Where R_3 is the outside radius of the pipe + insulation, R_2 the outside radius of the pipe, $k_{nsulation}$ the thermal conductivity of the insulation and h_{out} the heat transfer coefficient at the outside insulation surface.

The NREL pretreatment process employs 3 screw-feed reactors each with a volumetric flow-rate of approximately 3.43×10^6 Lh^{-1} and an inside radius of 1.3 m. The thickness and insulation of the vessel were assumed to be 0.015 m and 0.1 m respectively resulting in1.315 m as the R_2 value and 1.415 m as R_3. The cladding material used for the reactor is Incoloy-825 and this at 158 °C, has a thermal conductivity value of about 13.8 Wm^{-1} °C^{-1} while the heat transfer coefficient of ambient air (h_{out}) is typically between 40 and 50 Wm^{-2} °C^{-1} depending on the climate and wind speed outside the vessel. Deciding on an average value of 45 Wm^{-2} °C^{-1} for h_{out}, the value of U was calculated to be 34 Wm^{-2} °C^{-1} and with a reactor length of 9 m, the heat loss per reactor was approximately 100 MJ. Taking into account the residence time of 5 min and assuming a heating up time of 15 min, the theoretical heat loss per kilogram at a mass fraction of 30% and an assumed bulk density of 0.08 kg L^{-1} was found to be 3.63 kJ kg^{-1}.

The third approach exploited a heat loss model developed by Briggs of Burton. In this model, various parameters such as the reactor diameter, temperature difference, insulation thermal conductivity and other factors had to be quantified; most of which were specified as the same values in the previous approach. The outcome of this approach resulted in a heat loss of 4.17 kJ kg^{-1} (See Appendix A for the description of the calculation method and a calculation example of one of the heat loss components along with the parameters used in calculating the other components). This value is comparable to the result obtained from the second approach (3.63 kJ kg^{-1}), indicating that the heat loss in this process is approximately 4 kJ kg^{-1}. The heat loss calculated in the first approach was significantly over estimated. However, compared to the 'heating' energy of the process, the amount of heat lost is of little significance.

Heat Recovery

In the development of the above energy consumption model, the theoretical energy required to hydrolyse each unit of biomass into hydrolysate was examined. On a commercial scale pretreatment process however, there is energy that can and should be recovered in order to reduce the net energy input. In this case, the energy that potentially could be recovered in Stages 3 and 5 (Fig. 1) was estimated by calculating the enthalpy changes of the components at each stage. Based on the corn stover composition, the percentage of glucose, xylose and other component in the hydrolysate at Stages 3 and 5 were modelled as 35.5%, 20.0% and 44.5%, respectively. And as 5% of the total water was assumed to have evaporated in stage 1, the amount of water at Stages 3 and 5 was modelled as 95% of the total inlet water mass – 2.20 kg kg^{-1} dry corn stover. The heat content for stages 3 and 5 was calculated using Equation (18);

$$E = E_{glucose} + E_{xylose} + E_{rem.substrate} + E_{water} \qquad (18)$$

Where the change in enthalpy for glucose and xylose was calculated using average values of their specific heat capacity values while that of the remaining substrate was calculated using Equation (9) and the change in enthalpy for water using the *Cp* equation in Equation (5). The average molar specific heat capacity values for glucose and xylose were found to be 224 J mol^{-1} K^{-1} and 184 J mol^{-1} K^{-1} respectively [26]. Equation (18) can therefore be re-written as;

$$\Delta E = (n_{glu\,cose} \times C_{p_{glu\,cose}} \times \Delta T) + (n_{glu\,cose} \times C_{p_{xylose}} \times \Delta T)$$

$$+ m_{rem.substrate} \int_{T_1}^{T_2} (0.00004T^2 - 0.0015T + 0.9325)dT$$

$$+ mH_2o \int_{T_1}^{T_2} (0.00001T^2 - 0.0013T + 4.2085)dT \tag{19}$$

Where the specific heat capacity of the hydrolysate was worked out within the temperature limits of the system 130 °C and 158 °C for stage 3. The result of the calculations was approximately 299.0 kJ kg^{-1}. The same approach was used for Stage 5 from 97 °C to 130 °C and the outcome was 345.7 kJ kg^{-1}. The total amount of energy that could be given off was therefore found to be 644.7 kJ kg^{-1}.

Of this amount, the total amount that a heat exchanger is capable of recovering is between 60% and 70% [27]. Assuming 65%, the total energy recovered from the process is therefore estimated to be 419.1 kJ kg^{-1}.

The heat that potentially can be recovered from the cooling stages in the pretreatment process is about 25% of the calculated energy required. The heat recovered is believed to be of a lower quality than that of the steam injected into the system. However, the recovered heat can be used to heat the reactants up as far as is possible or to heat the enzyme hydrolysis tank. Nevertheless, the exact amount of energy that could be recovered depends on the efficiency of heat exchange system, the pretreatment operating parameters and the degree of process integration.

SENSITIVITY ANALYSIS

Five key variables were considered for the sensitivity analysis; the percentage of water evaporated, the solids loading rate, the acid loading, the operating temperature and the reaction conversion percentages. Reasonable ranges for the variables were selected based on estimates of the most probable ranges obtained from literature, with the baseline values for the variables set as the figures used in developing the model as depicted in Table 3.

Table 3: Raw data of the dilute acid pretreatment energy sensitivity analysis

Variables	Value	Range
Water evaporated/%	5	0–100 (increments of 5)
Solids loading rate/%	30	5–30 (increments of 5)
Acid loading/mg g^{-1}	18	10–28 (increments of 1)
T/°C	158	120–190 (increments of 5)
Reaction conversions	NREL conversion values for the 18 reactions displayed in Table 1 and Table 3	20 different conditions for the reactions were obtained from literature. (see Appendix B)

Fig. 3a shows that the solids loading rate is the most sensitive variable to the model followed by the percentage of water evaporated while the operating temperature has a moderate impact and the acid loading together with the reaction yield the least impact.

The solids loading rate and the percentage of water evaporated were particularly sensitive, for example, when the percentage of water evaporated was increased from 5% (base case) to 100% (see Fig. 3a[i]) and the solids loading rate decreased from 30% (base case) to 5% (see Fig. 3a [ii]), the total pretreatment energy per kilogram increased by approximately 274% and 658%, respectively. This further reiterates the conclusion made about the solids loading rate and as such, optimization should be on achieving a pretreatment reactor that can accommodate high solids loading in order to minimize the amount of water used and hence the energy consumed. Second generation ethanol pretreatment facilities operating at higher solids will require less energy in the form of heat to raise the temperature of the feedstock for pretreatment. However, increasing the total solids of the feedstock will affect the size of the motor required to mix and transport the feedstock. There are limited papers however citing solid loading rates greater than 30%, as above 30% solids loading rate, the wet material is likely to be un-pumpable in practice.

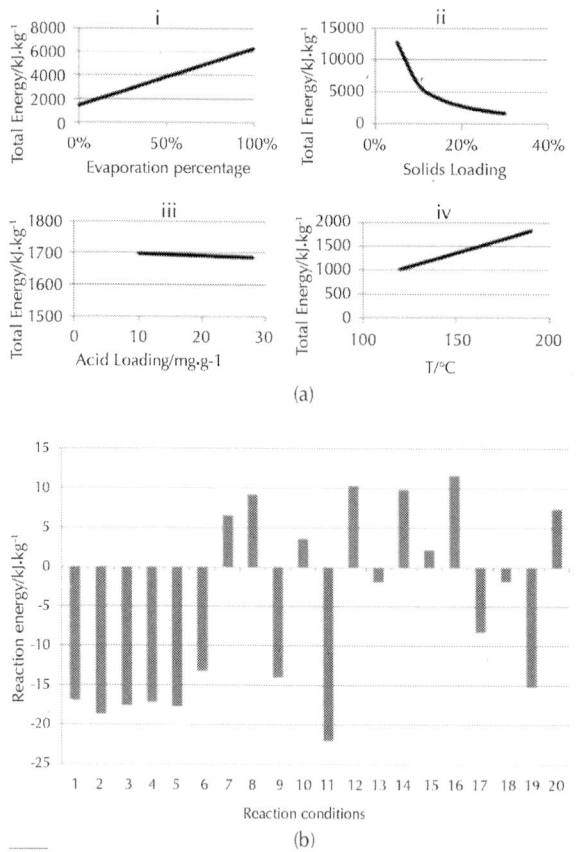

Figure 3: (a) Effect of the percentage of water evaporated (i), solids loading rate (ii), acid loading (iii) and operating temperature (iv) on the total pretreatment energy (kJ). (B) Effect of various reaction yields on the reaction energy.

On the other hand, most pretreatment processes reported in literature are carried out at a solids loading rate of 10% which coincides with the inflexion point in Fig. 3a(ii). These research works are however laboratory studies carried out for the purpose of investigating the impact of different conditions where low solids loading rate help to provide the greater mass and energy transfers needed for high hydrolysis yields. Miscibility is another factor as mixing in the laboratory is a lot more efficient at lower solids loading rates. The amount of energy consumed during pretreatment is of little or no concern in such studies. Chen et al. [28] has however quoted a solids loading rate of 45% for a dilute acid pretreatment process using corn stover at a temperature of 150 °C (for 5, 10 and 20 min) and an acid loading per gram of biomass of about 8 mg.

The effect of the acid concentration on the total energy consumed shows that it was not a major contributor to the energy efficiency of the pretreatment process (Fig. 3a [iii]). Varying the acid concentration per gram of dry biomass from 10 mg to 28 mg (18 mg being the base case) showed only a marginal decrease in the energy consumed from 1698 kJ kg^{-1} to 1686 kJ kg^{-1}. It will be interesting however to examine the effect of acid concentration on the sugar yield.

In the case of the operating temperature, the range of 120 °C–190 °C was chosen based on a range of maxima and minima values obtained from several papers on dilute acid pretreatment processes [8], [20], [29], [30], [31] and [32]. Fig. 3a [iv] shows that an increase in the operating temperature results in an increase in the total pretreatment energy.

Analyses of the chemical reaction conversion rates (Table 1) were conducted to assess its effects on the reaction energy; a constituent of the total pretreatment energy. 19 dilute acid pretreatment reaction conversions other than the NREL report were obtained from literature [20] and [33] (See Appendix B). In incidences where reaction conversions were not specified they were assumed to be the same as the base case scenario. Fig. 3b shows a variation in the reaction energy outcome from negative to positive values proving that the reaction conversions significantly impact the reaction energy. However, as the overall reaction energy was still insignificant compared to the heating energy and thus the total energy, the reaction conversions factor was not considered to be an important one for the optimization of the pretreatment energy efficiency.

MODEL VERIFICATION

Verification Approach 1

The mathematical model proposed for the energy estimation of dilute acid pretreatment processes was tested against the data provided by the NREL dilute acid pretreatment process. The process had a daily capacity of 2000 t with an expected up time of 96%. The numerical results of the model were compared to the NREL benchmark figures in order to verify the model's accurate representation of the pretreatment process description and solutions.

The NREL pretreatment process description provides a comprehensive review with all the flow-rates and temperature values and was used in the verification process. In the NREL's Process Flow Diagram (PFD)[21], there are five input streams to the pretreatment vessel; the milled corn stover stream, the process water stream, the sulphuric acid stream and the two high pressure steam streams. The high pressure steam is injected into the vessel to obtain and maintain the reaction temperature and as such, the amount of energy carried by these streams was calculated using a generic form of the model developed, and represented in Equation (20). Furthermore, the amount of energy required by the corn stover, sulphuric acid and process water were calculated.

$$\Delta E = m \overline{C}_p \Delta T \tag{20}$$

In these calculations however, the initial temperature of the components were those specified in the NREL report as well as the mass flow-rates. The flow-rates and temperature values for each stream can be found inTable 4a. The model includes the following average specific heat capacities: corn stover (1.03 kJ kg^{-1} °C^{-1}), process water (4.2 kJ kg^{-1} °C^{-1}), high pressure steam (2.8 kJ kg^{-1} °C^{-1}) and sulphuric acid (1.34 kJ kg^{-1} °C^{-1}).

Table 4: (a) Conditions of the NREL pretreatment process used in the verification process. (b) Range of pretreatment energies at varying water evaporation rates

a)

Component	Units	Corn stover	Process water	Sulphuric acid	H.P Steam	H.P Steam
Total flow rate	kg h⁻¹	104,167	140,850	38,801	3490	24,534
Insoluble solids	%	67.7	0	0	0	0
Soluble solids	%	12.3	0	0	0	0
Temperature	°C	25	95	113	268	268
Pressure	kPa	101	476	618	1317	1317
Water mass flow rate	kg h⁻¹	20 833	140 850	36 767	3490	24 534

b)

	Units	Water evaporation rate					
		0%	1%	2%	3%	4%	5%
Model pretreatment energy	kJ kg⁻¹	1434	1482	1531	1579	1628	1676
Dry corn stover mass flow rate	kg h⁻¹	83,334	83,334	83,334	83,334	83,334	83,334
Total pretreatment energy	GJ h⁻¹	120	124	128	132	136	140

Both the corn stover and sulphuric acid streams contain water and as a consequence were calculated separately from the corn stover and sulphuric acid components. The energy required per kilogram of dry corn stover was calculated to be approximately 809 kJ.

Similarly, the amount of energy per kilogram of dry corn stover carried and supplied by the two steam streams with flowrates of 3490 kg h^{-1} and 24 534 kg h^{-1} were calculated from 268 °C to 158 °C; taking into account also the latent heat energy. The finding of these calculations was approximately 808 kJ. This is similar to the results obtained by the energy required, suggesting that the model is reliable with the data used in construction.

Verification Approach 2

In another approach, the numerical results of the model the total pretreatment energy on a kilogram basis (kJ kg^{-1}) was multiplied by the mass flow-rate of the dry corn stover to attain the total pretreatment energy for the NREL pretreatment process. This calculation was executed a number of times using a range of pretreatment energies (kJ kg^{-1}) obtained from the sensitivity analysis of the percentage of water evaporated. The pretreatment energy ranged from 0% to 25% of water evaporated in increments of 1%.

The energy supplied to the process was gathered from various high energy content streams, such as high temperature process water and the water in the acid stream in addition to the two high pressure steams. This amount of energy was calculated and then subsequently used to determine at which water evaporation rate the energy supplied would meet the energy required As the results from the energy estimation model were used in determining the total energy required, the calculations of the total energy supplied were based on having the assumed initial temperature value of 25 °C as the final temperature and the stream temperatures as the initial temperature values.

A simplified form of Equation (7) was used in the steam calculations using average specific heat capacity values of 4.2 kJ kg^{-1} °C^{-1} and 2.8 kJ kg^{-1} °C^{-1} for water and steam, respectively. The calculations produced a total supplied energy of 138 GJ h^{-1} corresponding with the energy required by the pretreatment process using a water evaporation rate of 4% as shown in Table 4b.

Both verification approaches confirm that the majority of assumptions used to form the model can be validated.

APPLICATION OF THE MODEL TO OTHER DILUTE ACID PRETREATMENT PROCESSES

The developed energy estimation model was applied to several other dilute acid pretreatment processes with varying temperatures, pressures, solids loading rates, residence times, and acid concentrations. The biomasses used in the processes comprised of tall fescue, corn stover, switchgrass, rye straw, bermudagrass and wheat straw [20], [21], [29], [30], [31] and [32].

The total pretreatment energies of these processes were obtained using similar assumptions as applied in the original model together with new assumptions e.g. the selection of specific heat capacity values for the various biomasses. These energy values were then plotted with their respective glucose and xylose yields as presented in Fig. 4a and b respectively. Out of the 80+ dilute acid pretreatment conditions modelled, roughly only 40 reported the glucose sugar yields directly after the pretreatment process. Nonetheless, the scatter of the glucose and xylose yield data points in relation to the pretreatment energy suggests that there is no direct link between the percentage of glucose and xylose sugars released and the total pretreatment energy required for dilute acid pretreatment processes. Such plots can therefore be used to eliminate processes that are not economically viable. Nevertheless, the lack of correlation of the sugar yield and pretreatment energy could be attributed to a number of factors that were not taken into account such as; the type of biomass used and its state, that is, whether it was dried and/or milled, the way the sugars are structured in the biomass and how accessible they are. The lignin percentage in the raw material may also be a factor to consider in the pretreatment –glucose/xylose yield relationship.

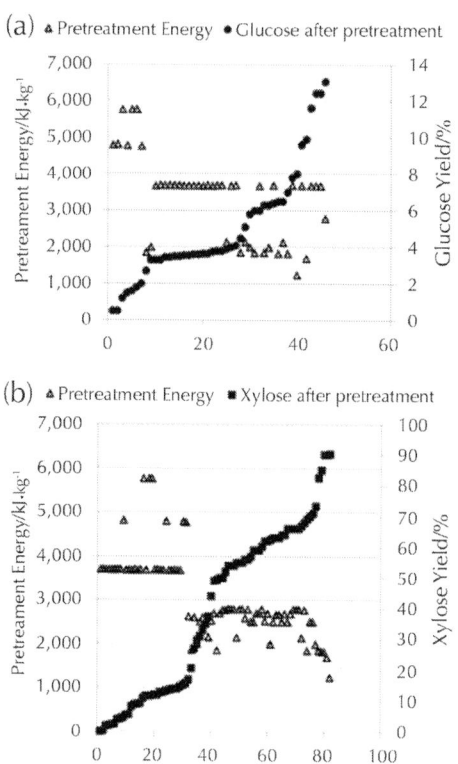

Figure 4: (a) Relationship between various dilute acid pretreatment energies and their glucose yields. (b) Relationship between various dilute acid pretreatment energies and their xylose yields.

CONCLUSIONS

An energy estimation model was developed to evaluate the energy demand in dilute acid pretreatment processes. The majority of the energy required was found to be from the heating stage of biomass for the pretreatment reaction. In comparison the reaction energy and maintenance energy were insignificant. For the process reported by NREL, 1692 kJ kg^{-1} was required to pretreat the material. Solid loading rate was found to be a key factor in influencing the energy requirements during pretreatment. Reducing the amount of process water increases the concentration of material in the process, which could potentially increase both the concentration of sugars liberated and reduce the energy required to heat the water to the reaction temperature. The heat recovery in the cooling steps could reduce the net energy requirement significantly. This model could be used as a decision-making tool for pretreatment selection, design and process optimization.

ACKNOWLEDGEMENTS

The research reported here was supported (in full or in part) by the Biotechnology and Biological Sciences Research Council (BBSRC) Sustainable Bioenergy Centre under the programme for 'Lignocellulosic Conversion to Ethanol' (LACE) (grant BB/G01616X/1). This is a large interdisciplinary programme, and the views expressed in this paper are those of the authors alone and do not necessarily reflect the views of the collaborators or the policies of the funding bodies. Oluwakemi Mafe is supported by a CASE PhD studentship funded by the BBSRC and Briggs of Burton plc. Scott Davies and John Hancock are employed by Briggs of Burton plc. The author would also like to acknowledge all the work and generous help from the National Renewable Energy Laboratory (NREL).

REFERENCES

1. The World Bank Group [Internet] Energy – the facts [cited 2013 Jul 16] DC (USA), Washington (2013) Available from: http://web. worldbank.org/WBSITE/EXTERNAL/TOPICS/EXTENERGY2/0,co ntentMDK:22855502~pagePK:210058~piPK:210062~theSite PK:4114200,00.html

2. R.G. Candido, G.G. Godoy, A.R. Goncalves Study of sugarcane bagasse pretreatment with sulphuric acid as a step of cellulose obtaining World Acad Sci Eng Tech, 61 (2012), pp. 101–105.

3. P. Bondesson, M. Galbe, G. Zacchi Ethanol and biogas production after steam pretreatment of corn stover with or without addition of sulphuric acid Biotechnol Biofuels, 6 (2013), pp. 1–11

4. N. Pensupa, M. Jin, M. Kokolski, D. Archer, C. Du A solid state fungal fermentation-based strategy for the hydrolysis of wheat straw Bioresour Technol, 149 (100) (2013), pp. 261–267

5.   S. Burkhardt, L. Kumar, R. Chandra, J. Saddler How effective are traditional methods of compositional analysis in proving an accurate material balance for a range of softwood derived residues? Biotechnol Biofuels, 6 (2013), pp. 1–10

6.   M. Balat, H. Balat, C. Oz Progress in bioethanol processing Progr Energ Combust Sci, 34 (2008), pp. 551–573

7.   S. Kim, B.E. Dale Global potential bioethanol production from wasted crops and crop residues Biomass Bioenerg, 26 (2004), pp. 361–375

8.   B. Yang, C.E. Wyman Pretreatment: the key to unlocking low-cost cellulosic ethanol Biofuels Bioprod Biorefin, 2 (2008), pp. 26–40

9.   P. Kumar, D.M. Barrett, M.J. Delwiche, P. Stroeve Methods for pretreatment of lignocellulosic biomass for efficient hydrolysis and biofuel production Ind Eng Chem Res, 48 (2009), pp. 3713–3729

10.   F.M. Girio, C. Fonseca, F. Carvalheiro, L.C. Duarte, S. Marques, R. Bogel-Lukasik Hemicelluloses for fuel ethanol: a review Bioresour Technol, 101 (2010), pp. 4775–4800

11.   L. Sousa, S.P.S. Chundawat, V. Balan, B.E. Dale 'Cradle-to-grave' assessment of existing lignocellulose pretreatment technologies Curr Opin Biotechnol, 20 (2009), pp. 339–347

12.   M. Balat Production of bioethanol from lignocellulosic materials via the biochemical pathway: a review Energ Convers Manag, 52 (2011), pp. 858–875

13.   O.A.T. Mafe, N. Pensupa, E.M. Roberts, C. Du Generation of bioenergy

14.   C. Lin, R. Luque (Eds.), Renewable resources for Biorefineries, The Royal Society of Chemistry, Cambridge (2014), pp. 117–145

15.   T. Eggeman, R.T. Elander Process and economic analysis of pretreatment technologies Bioresour Technol, 96 (2005), pp. 2019–2025

16.   D. Klein-Marcuschamer, P. Oleskowicz-Popiel, B.A. Simmons, H.W. Blanch Technoeconomic analysis of biofuels: a wiki-based platform for lignocellulosic biorefineries Biomass Bioenerg (2010), pp. 1–8

17.   J. Littlewood, R.J. Murphy, L. Wang Importance of policy support and feedstock prices on economic feasibility of bioethanol

production from wheat straw in the UK Renew Sustain Energ Rev, 17 (2013), pp. 291–300

18. L. Tao, A. Aden, R.T. Elander, V.R. Pallapolu, Y.Y. Lee, R.J. Garlock, *et al.* Process and technoeconomic analysis of leading pretreatment technologies for lignocellulosic ethanol production using switchgrass Bioresour Technol, 102 (2011), pp. 11105–11114

19. L. Wang, J. Littlewood, R.J. Murphy Environmental sustainability of bioethanol production from wheat straw in the UK Renew Sustain Energ Rev, 28 (2013), pp. 715–725

20. C. Conde-Mejia, A. Jimenez-Gutierrez, M. El-Halwagi A comparison of pretreatment methods for bioethanol production from lignocellulosic materials Process Saf Environ Prot, 90 (2012), pp. 189–202

21. D. Kumar, G.S. Murthy Impact of pretreatment and downstream processing technologies on economics and energy in cellulosic ethanol production Biotechnol Biofuels, 4 (2011), pp. 1–19

22. D. Humbird, R. Davis, L. Tao, C. Kinchin, D. Hsu, A. Aden, *et al.* Process design and economics for biochemical conversion of lignocellulosic biomass to ethanol: dilute-acid pretreatment and enzymatic hydrolysis of corn stover

23. National Renewable Energy Laboratory, Office of Energy Efficiency and Renewable Energy, Colorado (2011 May) Report No.: NREL/TP-5100-47764. Contract No.: DE-AC36-08GO28308. Sponsored by the US Department of Energy

24. The Engineering Toolbox [Internet] [Place unknown]: water – thermal properties (2013) [cited 2013 Sep 2]. Available from: http://www.engineeringtoolbox.com/water-thermal-properties-d_162.html

25. The Engineering Toolbox [Internet] [Place unknown]: water vapor – specific heat (2013) [cited 2013 Sep 9]. Available from: http://www.engineeringtoolbox.com/water-vapor-d_979.html

26. The Engineering Toolbox [Internet] [Place unknown]: liquids and fluids – specific heats (2013) [cited 2013 Sep 21]. Available from:http://www.engineeringtoolbox.com/specific-heat-fluids-d_151.html

27. Sulphuric Acid (H_2SO_4) Technical bulletin [Internet]. Canada Colors & Chemicals Limited; [updated 2002 March; cited 2014 Jan 13]. Available from: http://www.ccc-group.com/uploads/CCC%20Process%20Diagrams/Acid_Electronic_copy.pdf.

28. F. Kawaizumi, N. Sasaki, Y. Ohtsuka, H. Nomura, Y. Miyahara The specific heat capacities of d-glucose and d-xylose in water-ethanol system Bull Chem Soc Jpn, 57 (11) (1984), pp. 3258–3261

29. L. Theodore Heat transfer applications for the practicing engineer John Wiley & Sons Inc., New Jersey (2011)

30. X. Chen, J. Shekiro, M.A. Franden, W. Wang, M. Zhang, E. Kuhn, et al. The impacts of deacetylation prior to dilute acid pretreatment on the bioethanol process Biotechnol Biofuels, 5 (2012), pp. 1–14

31. Y. Sun, J.J. Cheng Dilute acid pretreatment of rye straw and bermudagrass for ethanol production Bioresour Technol, 96 (2005), pp. 1599–1606

32. B.C. Saha, L.B. Iten, M.A. Cotta, Y.V. Wu Dilute acid pretreatment, enzymatic saccharification and fermentation of wheat straw to ethanol Process Biochem, 40 (2005), pp. 3693–3700

33. Y. Zhu, Y.Y. Lee, R.T. Elander Optimization of dilute-acid pretreatment of corn stover using a high-solids percolation reactor Appl Biochem Biotechnol, 121–124 (2005), pp. 1045–1054

34. D.J. Schell, J. Farmer, M. Newman, J.D. McMillan Dilute-sulfuric acid pretreatment of corn stover in pilot-scale reactor Appl Biochem Biotechnol, 105-108 (2003), pp. 69–85

35. A.M.J. Koostra, H.H. Beeftink, E.L. Scott, J.P.M. Sanders Optimization of the dilute maleic acid pretreatment of wheat straw Biotechnol Biofuels, 2 (31) (2009), pp. 1–14

Increasing Biomass Resource Availability Through Supply Chain Analysis

Andrew Welfle, Paul Gilbert[1], and Patricia Thornley[2]

[1]Tyndall Centre for Climate Change Research, School of Mechanical, Aerospace and Civil Engineering,

[2]University of Manchester, Pariser Building, Sackville Street, Manchester M13 9PL, United Kingdom

ABSTRACT

Increased inclusion of biomass in energy strategies all over the world means that greater mobilisation of biomass resources will be required to meet demand. Strategies of many EU countries assume the future use of non-EU sourced biomass. An increasing number of studies call for the UK to consider alternative options, principally to better utilise indigenous resources. This research identifies the indigenous biomass resources that demonstrate the greatest promise for the UK bioenergy sector and evaluates the extent that different supply chain drivers influence resource availability.

The analysis finds that the UK's resources with greatest primary bioenergy potential are household wastes (>115 TWh by 2050), energy crops (>100 TWh by 2050) and agricultural residues (>80 TWh by 2050). The availability of biomass waste resources was found to demonstrate great promise for the bioenergy sector, although are highly susceptible to influences, most notably by the focus of adopted waste management strategies. Biomass residue resources were found to be the resource category least susceptible to influence, with relatively high near-term availability that is forecast to increase – therefore representing a potentially robust resource for the bioenergy sector. The near-term availability of UK energy crops was found to be much less significant compared to other resource categories. Energy crops represent long-term potential for the bioenergy sector, although achieving higher limits of availability will be dependent on the successful management of key influencing drivers. The research highlights that the availability of indigenous resources is largely influenced by a few key drivers, this contradicting area of consensus of current UK bioenergy policy.

INTRODUCTION

The UK energy sector is facing it's greatest challenges for at least a generation. The sector is expected to renew its energy generation portfolio, whilst providing secure, reliable, affordable and low carbon energy to its customers [1]. Energy from biomass provides options for the energy sector that can provide parts of the solution to each of these challenges.

Despite some concerns over the extent of biofuels deployment, bioenergy is key to many European energy strategies [2]; the European Commission estimates that energy from biomass may contribute up to two-thirds of the EU's 2020 target for 20% renewable energy contribution [3]. The UK's Renewable Energy Strategy also confirms that energy from biomass will significantly contribute towards the UK's energy portfolio [4].

Inclusion of energy from biomass pathways in both national and global energy strategies [5] means that increased mobilisation of biomass resource will be required to meet demand. The energy strategies of many EU countries currently assume the extensive use of non-EU sourced biomass [6], which will increase competition for suitable feedstocks [7].

The UK faces urgent choices regarding the future direction of its bioenergy sector. Ifcurrent plans mature the sector will be increasingly dominated by large scale biopower co-firing systems that will lock the UK into indigenous deficits of the feedstocks required to keep these plants running [8] and [9]. There are an increasing number of studies and calls [9], [10], [11], [12] and [13] for the UK to consider alternative biomass options, principally to make better use of the indigenous resources available. Welfle et al. [9] showed, through the development of a series of UK biomass resource scenarios, that the UK could potentially deliver 22% of its primary energy demand in 2050 through indigenously grown biomass and energy crops, 6.5% through the utilisation of indigenous residue resources from ongoing activities and a further 15.4% from waste resources.

The UK has many potential sources of biomass suitable for energy options. If indigenous resources are to be increasingly utilised, it is important that a greater understanding is achieved, of how different influencing drivers determine the extent that biomass resources become available to the bioenergy sector. Assessing the availability of any given resource being a matter of evaluating how much it is realistically, environmentally and economically viable to be made available to the energy market [14]. Some of these key drivers can be categorised as follows [15] and [16]:

- Policy Drivers – energy and environmental themed policies are particularly important in determining a secure long-term energy strategy. Waste, agricultural and forestry policies have great influence in determining the potential availability of specific resources.

- Market Drivers – biomass is a relatively immature market in the UK. The level of understanding of the UK biomass resource markets, determines the levels of uncertainty and ultimately the likelihood of commitments to long term bioenergy contracts.

- Technical Drivers – are the influences and barriers that may influence the actually processes of energy generation. These may include issues such as the availability of fuel standards or the ability to integrate biomass resources with the existing fossil fuel dominated network.

- Infrastructure Drivers – influences relating to the performance of all facilities required for the bioenergy sector to operate, including the, harvesting, collection, storage and transport of feedstocks.

The aims of this Paper are to identify and evaluate the most significant drivers within supply chains that influence the availability of UK indigenous biomass resources for potential utilisation by the bioenergy sector.

The objective is to inform the developers of bioenergy strategy and policy, and the wider bioenergy sector of opportunities to increase biomass resource availability. This is enabled through: highlighting specific indigenous resources that represent robust and continuous options for the bioenergy sector; identifying specific supply chain drivers that are found to command the greatest influence in determining the availability of biomass resources; identification of the resources whose availabilities are found to be most and least susceptible to variances within supply chains; and highlighting areas where policy measures should potentially focus in order to maximise the availability of indigenous resource.

Although this research is focused on the UK, the analysis is also applicable to similar case studies where a greater understanding of indigenous biomass availability is sought.

The research analysis is undertaking utilising a Biomass Resource Model (BRM), developed to simulate the whole system dynamics of biomass supply chains. The BRM brings together and allows the calibration of a wide range of drivers and variables that collectively determine the potential indigenous resource availability to 2050. Within this research the BRM is utilised in undertaking a sensitivity analysis to evaluate the influence of how supply chain drivers influences the availability of different categories of biomass.

METHODOLOGY

The following section introduces and discusses the analysis methodologies applied within this Paper. This includes an overview of the methodology developed within the BRM, and also that for measuring the extent that different supply drivers influence indigenous resource availability.

The Biomass Resource Model

The Biomass Resource Model is a resource focused modelling tool that enables the bottom up analysis of the practical potential of indigenous biomass resources, in this case within the UK. The drivers that control the BRM collectively reflect the variances and dynamics that influence biomass supply chains. Calibration of these drivers within the BRM allows the generation of realistic resource availability forecasts up to 2050. These drivers are discussed further and listed in Section 3 of this Paper.

A summary of the BRM's high level design is shown in Fig. 1. This highlights that the BRM's analysis methodology progresses in three distinct stages as described below. A greater depth discussion of the BRM's methodologies including an overview of the key research influences are described by Welfle et al.[9] and [17].

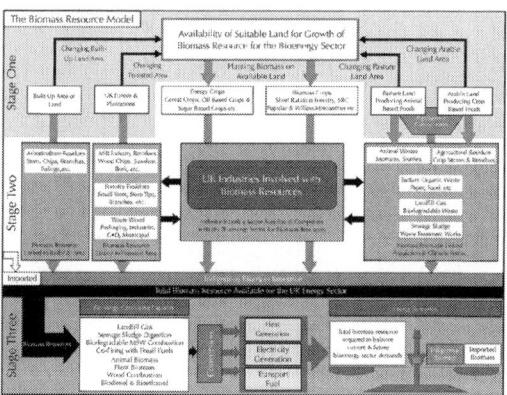

Figure 1: The Biomass Resource Model methodology architecture.

BRM Analysis Stage One: Land Use & Availability Analysis

Analysis stage one evaluates the area of UK land utilised to meet various demands, including; food production, further urban development and forestry to 2050. The remaining UK land area potentially suitable for crop production is then analysed to determine the potential availability for biomass and energy crop growth dedicated for the bioenergy sector.

BRM Analysis Stage Two: Biomass Resource Availability

The second analysis stage quantifies and forecasts the extent, availability and competing markets for different biomass resources indigenous to the UK. This takes into consideration factors such as the potential for resource collection/harvest, changes in the levels of arisings linked to industrial activity and agricultural residue utilisation. The biomass categories and specific resources analysed within the BRM to reflect the range of UK resources are shown in Table 1. These also represent the biomass categories analysed within this paper.

Table 1: Summary of the analysed biomass Categories & specific resources

Categories	Biomass resources
Grown Resource *from UK Land*	Energy crops (food species) Cereal crops, oil crops, sugar crops
	Biomass crops (non-food species) Grasses, short rotation forestry & coppices, other forestry
Residues Resource *from UK Forestry, Industries & Processes*	Forestry residues
	Crop residues Straws
	Animal residues Manures & slurries
	Arboriculture arisings
	Industry residues Sawmill, pulpmill & industry residues
Waste Resource *from UK Industries & Processes*	Waste wood Packaging, industrial, construction, demolition, municipal
	Tertiary organic waste Household, commercial, industrial papers, cardboards, textiles, foods, organic & kitchen, garden etc.
	Sewage – waste treatment

BRM Analysis Stage Three: Indigenous Bioenergy Potential

The third analysis stage calculates the bioenergy potential of the specific resource quantities calculated within stage two. The range of pre-treatment and energy conversion pathways applicable to different types of biomass are considered. Resource bioenergy potentials are calculated taking account of the resource and energy efficiencies reflective of each bioenergy generation pathway. Once the energy potentials of the available resources have been calculated, these can then be compared against respective renewable energy and bioenergy targets.

In summary the key features of the BRM important to this analysis are the ability to investigate the different supply chain drivers that influence biomass resource availability. Also to evaluate food–fuel interfaces by simultaneously considering the land requirements for food production, biomass production and other uses.

Developing a Methodology for Analysing Influences to Biomass Resource Availability

This section describes the methodology developed for analysing the extent that different drivers influence resource availability. The aim was to undertake an assessment of the maximum practical availability of different indigenous resources to 2050, determine the drivers that most influence resource availability, evaluate the 'availability robustness' of each resource, and identify any notable trends through time.

Developing a Baseline

For each of the drivers discussed in Section 3 that control the BRM, a literature review was carried out to analyse how these currently stand in the UK, and to develop an idea of how these may change to 2050. A database was produced that collated the range of values that literature and studies forecast for these. This database was then analysed to develop a series of average or mean values for each of the drivers to 2050. These values therefore represent a 'literature informed' mean or 'baseline scenario' of how the UK's biomass supply chains may function to 2050. Calibrating the BRM to reflect this baseline enabled an evaluation of the 'average' availability and bioenergy potential of each indigenous resource to 2050. Identifying which resources may be most abundant, and which resources may provide the greatest bioenergy potential being fundamental to this analysis.

Evaluating How Different Drivers Influence Biomass Resource Availability

The key element of the Paper's analysis is evaluating the extent that each supply chain driver influences resource availability. A sensitivity analysis methodology was developed so that the influence of each driver could be analysed in isolation of the others. This was achieved through calibrating the BRM to reflect the baseline scenario (discussed in Section The BRM was then progressively run to reflect the performance range of variances forecast by literature for each individual driver, whilst keeping all the other drivers set at the baseline. Undertaking this methodology for all drivers allowed an assessment to the extent that each influenced the availability of different biomass resource to 2050.

DRIVERS INFLUENCING BIOMASS AVAILABILITY

This next section introduces and provides further context to the supply chain drivers that make up the BRM. It also provides a review of selected literature and discussions for how they are deemed to influence the availability of different biomass resources in the UK.

All biomass resource models and assessments revolve around analysing the influence of different drivers. As such the range of drivers listed within literature that are identified as being influential of biomass resource availability is extremely broad. Table 2 presents an overview of many of these and highlights the capabilities and limitations of the BRM in analysing each.

Table 2: Summary of supply chain drivers that influence biomass resource availability

Categories	Supply chain drivers	References	BRM analysis capability
Economic & Development Drivers	Population change	[18], [19], [20], [21], [22], [23],[24] and [25]	✔
	Resource import/export	[18] and [23]	✔
	Economic & technical development	[20] and [26]	–
	Industry productivity	[20], [23] and [26]	✔
	Gross domestic product	[19], [24] and [25]	–
	Rural economy development	[27] and [28]	X
Infrastructure Targets	Energy system structure	[23], [27], [28] and [29]	✔
	Energy generation plant	[27] and [28]	✔
	Supply chain development	[27] and [28]	✔
Physical & Climate Drivers	Land use change	[20], [22], [23], [24] and [25]	✔
	Water availability	[21] and [25]	X
	Climate change	[18], [20], [21] and [25]	–
	Flood protection land requirements	[18]	X
	Nature conservation land requirements	[18] and [21]	–
	Soil degradation	[18] and [21]	X
Food Drivers	Per-capita food demand & consumption	[18], [19] and [21]	✔
	Calorie consumption	[19]	X
	Diet change	[19]	X
	Agriculture productivity yields	[18], [19], [21] and [22]	✔
Resource Mobilisation Technical Drivers	Technological advances	[22], [23], [24], [25], [26] and [29]	✔

		Forest system productivity	[22], [23], [24], [25], [26] and [29]	✔
		Industry & process residue generation	[22], [23], [24], [25], [26] and [29]	✔
		Forestry residues collection	[22], [23], [24], [25], [26] and [29]	✔
Resource Demand Drivers		Resource use by industry	[18], [19], [22], [24], [25], [26] and [29]	✔
		Demand for round wood	[19], [22], [24], [25] and [29]	✔
		Demand for wood fuel	[22], [24], [25], [26] and [29]	✔
		Demand for other resources	[19], [22], [24], [25] and [29]	✔
Policy Drivers		Greenhouse gas emission targets	[23], [25], [27] and [29]	–
		Energy efficiency & consumption targets	[23], [24], [25], [27] and [29]	✔
		Renewable & bioenergy targets	[23], [24], [25], [27] and [29]	✔
		Fuel security drivers	[23] and [27]	✔
		Support policies & mechanisms	[23], [24], [25], [27], [28] and [29]	X
Key	✔	The BRM allows the analysis of these drivers in terms of their influence on biomass resource availability and bioenergy potential.		
	–	The BRM allows the analysis of partial aspects of these drivers. Or can provide an indirect evaluation the drivers influence on biomass resource availability and bioenergy potential.		
	X	The BRM current design and outputs do not allow the analysis of these drivers.		

From this list, a series of key supply chain drivers are identified from literature and form the basis of analysis within the BRM and this research. The drivers analysed within this research therefore represent a non-exhaustive reflection of all the drivers that may influence biomass resource within supply chains. These BRM drivers are listed and categorised within Table 3, and their respective influences on biomass resource availability are discussed below.

Table 3: Summary of key supply chain drivers analysed within the BRM

Category	Drivers
UK Development Drivers	1) Population change
	2) Changes in built-up land area
Food Production System Drivers	3) Crop & agriculture productivity
	4) Food waste generation
	5) Food commodity imports
	6) Food commodity exports
	7) Utilisation of agricultural wastes & residues
Forestry & Wood-based Industry Drivers	8) Forestry expansion & productivity
	9) Wood-based industry productivity
	10) Imports of forestry product
	11) Exports of forestry product
Biomass Residue & Waste Utilisation Drivers	12) Utilisation of forestry residues
	13) Utilisation of industrial residues
	14) Utilisation of arboriculture arisings
	15) Waste generation trends
	16) Waste management strategies
Biomass & Energy Crop Strategy Drivers	17) Land dedicated for energy crop growth

UK Development Drivers

Population Change

Population growth is the fundamental influence for all long term outlooks relating to food and agriculture [30]. The expected large increases in global food demand 2030–2050 are based on forecasts of increasing population [31]. Food and agricultural systems are closely linked to many biomass resource supply chains, therefore population is a driver with likely influence on biomass availability.

Within the BRM population forecasts reflect the United Nations Population Division's forecast variants for the UK [32].

Built-up Land Area

Urbanisation is a further driver that influences food and agriculture systems [33]. Changes in the extent of built-up land area directly influence the potential availability of biomass through reducing the area of land that could otherwise be dedicated for biomass production.

The BRM utilises forecasts of current and future built-up land areas for the UK, as developed within the MOSUS Project (Modelling Opportunities and Limits for Restructuring Europe towards Sustainability) [34].

Food Production System Drivers

Crop and Agriculture Productivity

The productivity of land and agricultural yields are important drivers that directly influence the production of biomass. Where crop yields can be increased, agricultural land may be freed for growth of biomass and energy crops [9]. Also where biomass and energy crop yields can be enhanced, more resource can be produced from the land available.

Improvements and variances in food and crop systems productivity result from the collective influence of a range of manageable and external inputs. The UK has great strength in crop science, including increasing understanding of responses to global climate change [35]. Mueller et al. [36] suggest that the 'yield gap' – the difference between attainable & actual yields, will continue to be reduced. Other forecasts suggest that yield increases of 70% by 2050 are possible for most crops through improved nutrient management, irrigation and productivity techniques [36] and [37].

Haberl et al. [30] found that Western European yields could experience further mean increases of >16% from CO_2 fertilisation by 2050, resulting from climate change forces (>2% without CO_2 fertilisation).

Although whilst the main northern hemisphere producers may experience favourable conditions from climate change in the next 40 years, regions where rising food demand is most pronounced will likely see production hindered. This may lead to a greater number of countries relying on fewer high latitude producers – increasing vulnerability to extreme weather events in these regions [38].

Current and forecast crop and agricultural yields analysed within the BRM reflects those documented in a wide range of studies and literature, including predicted climate change impacts [15], [16], [38], [39], [40],[41], [42], [43] and [44].

Food Waste Generation

Food waste influences the availability of biomass resource in multiple ways. Food waste itself is a plausible resource for bioenergy generation pathways. At the same time food waste is a factor that reduces the supply chain efficiency – the greater waste from the system, the more land is required to produce food commodity quantities to meet demand.

Research estimates that 25–50% of food produced is wasted along the supply chain [45], [46] and [47]. 50% of the UK's food waste comes from households, where at some point at least 60% of this waste could have been consumed [48]. The European Commission is targeting a 50% reduction in food wastes by 2020 [49], and the UK Government Office for Science suggests that halving food waste by 2050 may be equivalent to 25% of current productivity [50] and [51].

These waste reduction targets are considered in the analysis, through the BRM utilising a series of forecasts [39], [50], [51], [52] and [53], to quantify UK food waste trends.

Food Commodity Imports and Exports

Food commodity import and export trends are drivers that can influence biomass availability, as they contribute towards determining the area of UK land that is required to produce the food quantities to meet demand. Any land dedicated for food production is therefore unavailable for biomass or energy crop growth.

The majority of the UK's imports come from the EU, with the Common Agricultural Policy and EU Directives strongly influencing the shape of the UK food system [54]. The UK currently produces about half of the food it consumes, and is ~60% 'self-sufficient' [55]. The UK Government's stance is that it "sees no economic or environmental rationale for Government to set targets to raise UK output of particular food products in step with changes in global food demand" [54].

The analysis takes into consideration these stances of future food import/export trends, the BRM utilising data from a series of studies [39], [52] and [56] to reflect the UK's path.

Utilisation of Agricultural Wastes and Residues

Agricultural wastes and residues reflect a resource category with sizeable potential for the bioenergy sector[10]. Welfle et al. [9] found that this category of biomass resource could deliver up to 80 TWh of bioenergy by 2050.

The key drivers determining the availability of this resource for the bioenergy sector are the extent to which it is harvested/collected and the competition for the resource. The BRM analysis reflects the wide range of research and studies that forecast the extent and timeframes to which these resources could be utilised for energy generation: 20%–100% of total resource could be utilised, with typically half of this being available for the energy sector [57], [58], [59], [60], [61], [62], [63], [64] and [65]. The UK Department for Food & Rural Affairs (DEFRA) provides sustainability guidance on the extent that agricultural residues should be returned to the soil to protect and enhance soil and biodiversity (10% Lower Limit, 50% Higher Limit) [66].

Forestry and Wood-based Industry Drivers

Forestry Expansion and Productivity

The extent and productivity of forestry systems directly influences the availability of resources for the bioenergy sector. Forests provide energy generation opportunities either through specifically harvested resources, or via the collection of residues. Forests also provide indirect opportunities for the bioenergy sector through supplying resource to wood-based industries, that in turn produce wastes and residues that can be utilised by the bioenergy sector.

The BRM utilises the UK Forestry Commission's expansion and productivity forecasts [67], [68], [69], [70],[71], [72], [73] and [74].

Wood-Based Industry Productivity

The ongoing activities of wood-based industries produce wastes and residues that provide an opportunity for the bioenergy sector. At the same time wood-based industries require raw forestry products, of which it competes directly with the bioenergy sector for the lower grades of resource.

The BRM utilises existing data [56], [75] and [76] and forecasts [76] that predict the trends and directions that UK wood industries may take.

Imports and Exports of Forestry Product

Forestry product import and export trends can influence the availability of biomass resource through determining the extent that the indigenous forestry systems are utilised. Where imports are increased and exports are reduced, there will be less strain on indigenous forestry systems to produce the wood resource required to meet demand. This may in turn provide increased opportunities for the bioenergy sector. Likewise reduced imports and increased exports would have the counter influence, putting greater strain on indigenous forests.

The BRM again utilises existing data [56], [75] and [76] and forecasts [76] that predict the trends and directions that UK forestry products imports/exports may follow.

Biomass Residue & Waste Utilization Drivers

Utilization of Forestry Residues

Forestry residues represent an opportunity for the bioenergy sector that is currently un-utilised in the UK [58]. The availability extent of this resource is dependent on the proportion extracted from forestry systems and the proportions left in-situ to maintain the health of the habitat.

The BRM reflects the full range of residue extraction levels recommended by research and studies [29], [58],[77], [78] and [79], from 10% to as much as 100% by 2020 [58].

Forest certification standards set by the Forestry Stewardship Commission (FSC Criterion 5.3 & 6.3), Ministerial Conference on the Protection of Forests in Europe (MCPFE Criterion 2 & 3) and Programme for the Endorsement of Forest Certification (PEFC Criterion 4) all provide details for the minimisation of on-site harvesting and residue processing, maintenance of ecosystem health and function and protection of biodiversity [78].

Utilisation of Industrial Residues

Biomass residues from ongoing industrial processes represent a potential opportunity for the bioenergy sector [9]. The key drivers influencing the availability of this resource category are the extent to which it can be collected/processed, and productivity of the UK wood-based industry.

The BRM utilises data that reflects current and forecast productivity trends for the UK's wood-based industries [56], [75] and [76], and also forecasts of potential industry residue utilisation for energy [58],[80] and [81].

Utilization of Arboricultural Arising

UK Local Authorities and tree surgeons produce thousands of tonnes of arboriculture arisings. The majority of this is currently land-filled, stored for landscaping applications or burnt onsite. Although with correct processing, handling, grading and storing, these residues provide an opportunity for the bioenergy sector[82]. The key drivers determining resource availability are the extent to which the resource is harvested/ collected and the competition for the resource.

The BRM utilises forecasts from a series of research and studies that forecast that up to 100% of arboriculture arising could be utilised by the bioenergy sector [58], [76] and [77].

Waste Generation Trends & Waste Management Strategies

The potential availability of waste resources for the bioenergy sector is influenced by two key drivers: The amount of waste being generated, and the strategy implemented for how the waste is managed. Welfle et al.[9] found that there is both potentially high variability and availability of this resource, forecasts ranging from 1.8 to 130.7 Mt by 2050 dependent on the waste generation and management strategies.

The BRM utilises a series of data sets [48], [61], [83], [84], [85] and [86] that reflect the UK's current waste system, and applies DEFRA forecast scenarios [61] and [85] to analyse how the implementation of alternative waste strategies may influence potential availability for the bioenergy sector.

Biomass & Energy Crop Strategy Drivers

Land Dedicated for Energy Crop Growth

The area of land dedicated for biomass and energy crop growth is a fundamental driver in determining the potential availability of grown resource. Energy crops have an important role to play in helping to achieve the UK's renewable energy targets [66] and [87]. The UK Department for Energy & Climate Change (DECC) estimate that for the UK to meet these targets, approximately 3500 km^2 of land needs to be dedicated for energy crops – a large increase from the current 250 km^2 utilised. Although 3500 km^2 seems large it currently reflects <2% of UK agricultural land – an area that could be easily realised through farmers utilising un-used/marginal lands [66] and [87].

A large number of reports and studies estimate that varying amounts of the UK's >170,000 km^2 of agriculture land could be dedicated for biomass resource growth [88] and [89]. Potential land dedication estimates range from 3,500 to 10,000 km^2[15], [16], [35], [59], [90], [91], [92] and [93], whilst the theoretical maximum available land for short rotation coppices and Miscanthus without impacting food systems have been estimated to be between 9,300 and 36,300 km^2[66] and [87].

The European Environment Agency (EEA) also reported that between 8,000 and 34,000 km² of land could be released in the UK by 2030 by reform of the Common Agricultural Policy [92]. Fischer et al. [65] estimating that half of this released land would be former grassland.

The BRM takes into consideration these estimates when determining the proportion of free land to be dedicated for biomass resource growth.

RESULTS–UK BIOMASS AVAILABILITY & SUPPLY CHAIN INFLUENCES

The following section provides the results, presented in the form of figures and Supplementary tables. These document the availability and bioenergy potential of different biomass resources to 2050. They also outline the results of the sensitivity analysis that evaluates the extent that different supply chain drivers influence biomass resource availability.

The Potential Availability of UK Biomass for the Bioenergy Sector

Fig. 2 documents the results of the analysis undertaken to determine the maximum availability of each category of UK biomass (Table 1), when the BRM is calibrated to reflect the literature informed baseline scenario to 2050. The resources availabilities in the research are presented in million tonnes (Mt) of dry basis biomass resource. This analysis reflects the range of forecast supply chain characteristics for each driver (Table 3) as informed by literature. The trends represented document the resource potential if the most influential drivers are managed so that maximum levels of resource availability are achieved. Through highlighting the range in resource availability between 2015 and 2050, Fig. 2 also provides an indication of the extent of actions that may be required to achieve the higher level forecasts.

- UK 'Grown Resources' are shown to have relatively low availability in 2015 (>1.9 Mt), but this potentially increases by >1503% by 2050 (to >31 Mt).

- UK 'Residue Resources' in 2015 are shown to have availability of >11.7 Mt, potentially increasing by >152% by 2050 (to >29.7 Mt).

- UK 'Waste Resources' in 2015 are shown to have availability of >15.2 Mt, potentially increasing by >491% by 2050 (to >90.0 Mt).

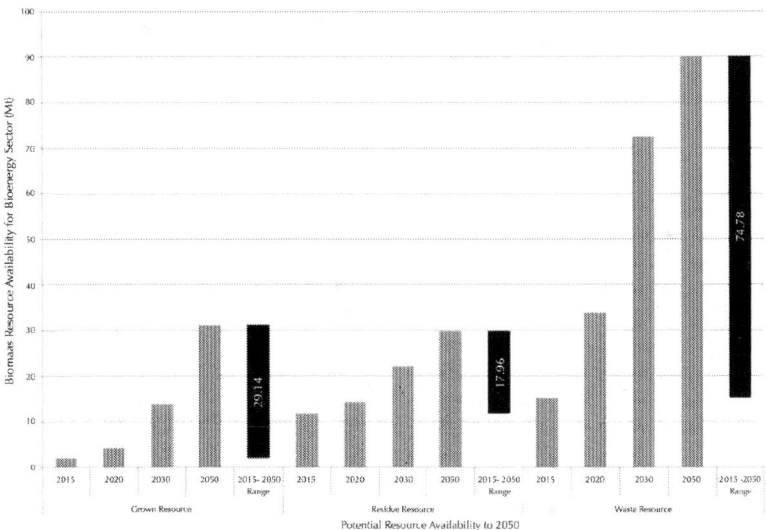

Figure 2: The potential availability of biomass resources for the bioenergy sector.

Analysing the Influence of Supply Chain Drivers on Biomass Resource Availability

Fig. 3, Fig. 4 and Fig. 5 present radar graphs that document the results of the supply chain driver sensitivity analyses. These show the extent that the different drivers influence resource availability, with each numbered spoke of the radar graphs reflecting the corresponding analysis for each of the numbered supply chain drivers (Table 3). The figures highlight the maximum availability of the each respective category of biomass resource to 2050 when the characteristics of each driver reflect the ranges informed by literature.

- Fig. 3 highlights that the key supply chain drivers influencing the availability of UK 'Grown Biomass Resources' are 'Population Change' (Driver 1), 'Crop & Agriculture Productivity' (Driver 3), 'Forestry Expansion & Productivity' (Driver 8) and 'Land Dedicated for Energy Crop Growth' (Driver 17).

- Fig. 4 highlights that the key supply chain drivers influencing the availability of UK 'Residue Biomass Resources' are 'Population Change' (Driver 1), 'Utilisation of Agricultural Wastes & Residues' (Driver 7) and the 'Forestry Expansion & Productivity' (Driver 8).

- Fig. 5 highlights that the key supply chain drivers influencing the availability of UK 'Waste Biomass Resources' are 'Waste Generation Trends' (Driver 15), and most notably by 'Waste Management Strategies' (Driver 16).

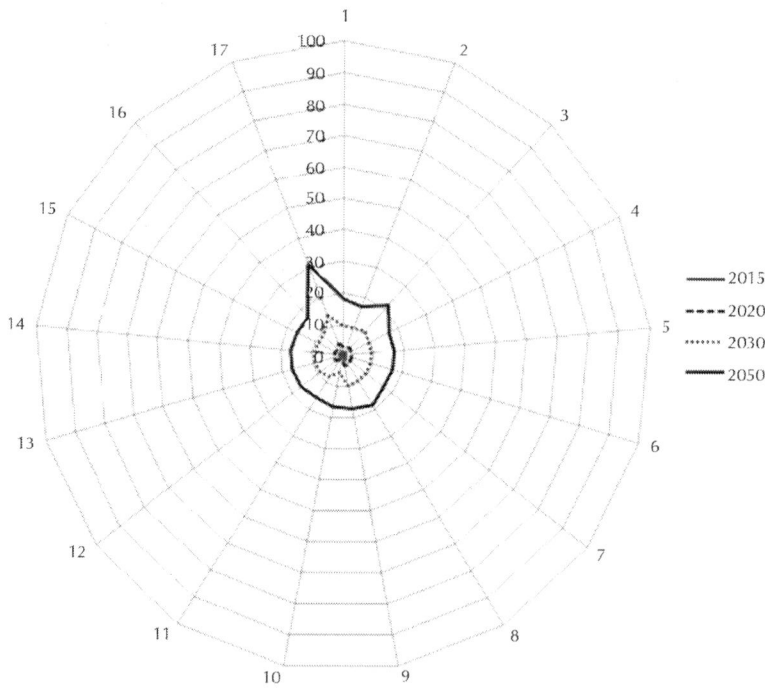

Figure 3: Analysis of drivers influencing the availability of grown biomass resources (Mt).

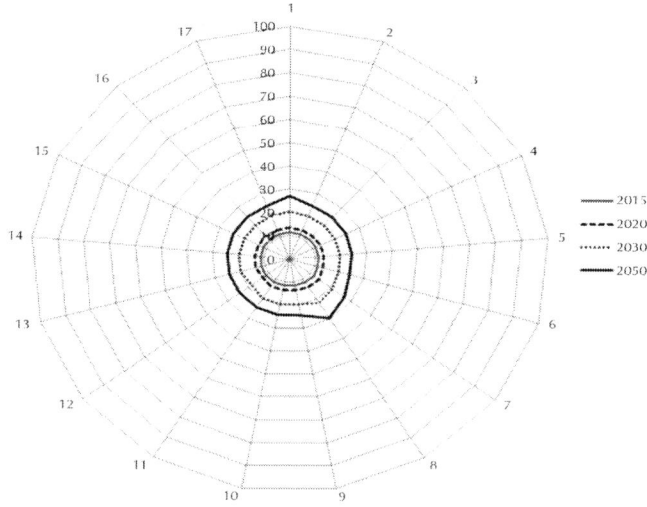

Figure 4: Analysis of drivers influencing the availability of residue biomass resources (Mt).

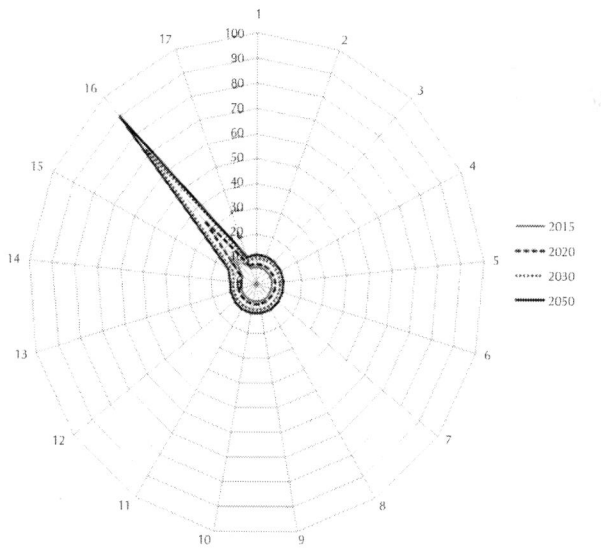

Figure 5: Analysis of drivers influencing the availability of waste biomass resources (Mt).

UK Resources Demonstrating the Greatest Potential for the Bioenergy Sector

Further analysis was carried out to determine which specific biomass resources may demonstrate the greatest potential for the bioenergy sector. The data from this analysis is documented within Appendix A1. This highlights the availability and bioenergy potential of different biomass resources, when supply chain characteristics reflect the literature informed baseline scenario to 2050. Further details describing the methodology for calculating bioenergy potential, including the applied conversion and pre-treatment pathways and efficiencies can be found in Welfle et al. [9] and [17]. From this analysis the following UK resources are shown to demonstrate particular availability for the bioenergy sector:

- UK 'Biomass & Energy Crops' from the Grown Resources Category (>31.2 Mt resource, equivalent to >104 TWh by 2050),
- UK 'Agricultural Residues' from the Residue Resources Category (>26.2 Mt resource, equivalent to >83 TWh by 2050),
- UK 'Household Wastes' from the Waste Resources Category (>40.7 Mt resource, equivalent to >117 TWh by 2050),
- UK 'Other Wastes' from the Waste Resources Category (>32.7 Mt resource, equivalent to >75 TWh by 2050).

Supply Chain Drivers with the Greatest Influence on Biomass Resource Availability

This section provides further discussion of the results of the sensitivity analysis. Evaluating the extent that the different drivers (Table 2) influence the biomass resources found to demonstrate the greatest potential availability in the UK: 'Biomass & Energy Crops', 'Agricultural Residues' and 'Household Wastes'. 'Other Wastes' are excluded from this further analysis, as this resource category represents a collection of all other wastes that are not classified as either 'Household' or 'Food or Organic' (Table 1). The data reflecting this analysis is included in Appendix A2.

Drivers Influencing the Availability of UK Biomass & Energy Crop Resources

Three drivers are shown to have significant influence in determining the availability of Biomass & Energy Crop resources. 'Population Change' (Driver 1) and 'Crop & Agricultural Productivity' (Driver 3) demonstrate marginal influence in determining the potential availability of this resource. However the 'Land Dedicated for Energy Crop Growth' (Driver 17) is shown to be the key influence. The results show that if the upper limits of land are made available for biomass and energy crop growth (Driver 17) as forecast by literature, the availability of this resource may be >87% greater in 2050 compared to scenarios where lower limits of land are utilised.

Drivers Influencing the Availability of UK Agricultural Residue Resources

The availability of agricultural residue resources is demonstrated to be influenced by 'Population Change' (Driver 1) and the 'Utilisation of Agricultural Residues' (Driver 7). The analysis shows that realisation of higher population forecasts (Driver 1) may potentially increase the availability of this resource in 2050 by >12.6%. Whilst realising upper limits of agricultural residue collection/harvests and utilisation (Driver 7) as forecast by literature, may result in >11.6% greater resource availability by 2050

Drivers Influencing the Availability of UK Household Waste Resources

The potential availability of household waste resources for the bioenergy sector is demonstrated to be influenced by both 'Waste Generation Trends' (Driver 15) and 'Waste Management Strategies' (Driver 16). Forecast trends of waste generation (Driver 15) are shown to have a minor influence on the availability of this resource. In contrast the results confirm that the waste management strategy adopted (Driver 16) represents a major influence. A waste management strategy complementing the bioenergy sector as forecast by literature may increase the availability of this resource: >318% by 2020, >476% by

2030 and >500% by 2050, compared to forecasts where waste is less utilised by the bioenergy sector.

DISCUSSION–MAXIMISING THE POTENTIAL OF UK BIOMASS

This next section provides a discussion of the results highlighted within Section 4. Identifying which of the UK's indigenous biomass resources may provide the best opportunities for the bioenergy sector, and how these relate to current UK policy.

The Potential of UK Biomass Resources for the Bioenergy Sector To 2050

For UK resources to substantially contribute towards meeting bioenergy targets, it is important to highlight which of the broad range of resources may provide the greatest potential and opportunities for the bioenergy sector.

The results presented within Fig. 2 represent three levels of analysis: the maximum potential availability of different categories of UK biomass; the extent that each resource category may be available in the near-term (by 2015); and also the range in potential resource increment between 2015 and 2050. The maximum availability potential is important, as it identifies how much resource could be mobilised for the bioenergy sector if influencing supply chain drivers are effectively managed. The near-term forecast and 2015–2050 increment ranges are important as they provide an insight into how much resource may be available without extensive further actions and management of drivers, and likewise provide an indication of the effort that may be required to achieve the higher levels of forecast availability.

Using this premise to evaluate the result for the three analysed biomass categories: Fig. 2 shows that UK 'Grown Biomass Resources' are forecast to have relatively low near-term availability, but large potential by 2050. This suggests that these resources may be highly influenced by supply chain drivers, and substantial effort may be required to manage these in order to increase the resource availability from the low base. Fig. 3confirms that the availability of land dedicated for the growth of these resources is the key driver requiring appropriate management if higher levels of resource availability are to be realised.

The research highlights that UK agricultural residues represent large resource opportunities for the bioenergy sector. Fig. 2 demonstrates that UK's 'Residue Biomass Resources' are shown to have 'medium' near-term availability compared to the other two resource categories. This increases at a steady rate to 2050 suggesting that residue resources are relatively robust to supply chain influences and less effort may be required to increase residue availability in comparison to the other resources categories. The relatively continuous increment in resource availability demonstrated by the spacing of the analysis time-lines withinFig. 4 also highlights that the availability of residue resources shows robustness to supply chain influences.

Household wastes are also found to represent large resource opportunities for the bioenergy sector. Fig. 2shows that UK 'Waste Biomass Resources' have near-term availability that exceeds the other two categories and the potential maximum increase in waste resource availability to 2050 is significant. This large increment suggests that waste resources are highly susceptible to supply chain influences, and significant effort may be required to manage these if the higher forecasts of resource availability are to be realised. This is reaffirmed within Fig. 5 where the influence of implemented waste management strategies is shown to be key. This research therefore highlights that in the long-term, wastes may represent resource options with significant potential for the bioenergy sector, albeit reliant on the implementation of complementary waste management strategies.

Increasing the Focus of Bioenergy Strategies

The UK Bioenergy Strategy aims to maximise the opportunities for improving the availability of all biomass resources through policies aimed at managing a broad range of supply chain drivers [87].

This research has analysed a wide range of supply chain drivers, finding large variances in their influence in determining biomass availability for the bioenergy sector. The research also highlights that particular resources demonstrate significantly greater availability and bioenergy potential than others. Therefore if the contribution of UK resources is to be maximised, the research suggests that bioenergy policies and strategies should become increasingly focused and targeted.

Table 4 summarises the research findings: ranking the UK's biomass resources based on their availability and bioenergy potential; also ranking the analysed supply chain drivers based on their influence in increasing UK biomass availability for the bioenergy sector.

Table 4: Analysis summary ranking UK biomass availability, bioenergy potential & supply chain influences.

Ranking	Influencing drivers	Resource availability & bioenergy potential for bioenergy sector
High Ranking *Drivers & resources with the greatest influence/ potential*	• Waste management strategies • Land dedicated for energy crop growth	• Agricultural residues • Household wastes • Biomass & energy crops • Other wastes
Medium Ranking *Drivers & resources with medium influence/ potential*	• Crop & agriculture productivity • Population change • Changes in built-up land area • Food waste generation • Utilisation of agricultural wastes & residues • Forestry expansion & productivity • Waste generation trends	• Dedicated forestry resources • Forestry residues • Food & organic wastes
Low Ranking *Drivers & resources with the least influence/ potential*	• Food commodity imports • Food commodity exports • Wood-based industry productivity • Imports of forestry product • Exports of forestry product • Utilisation of forestry residues • Utilisation of industrial residues • Utilisation of arboriculture arisings	• Sewage wastes • Industry residues • Arboricultural residues

Potential Strategies for Increasing UK Resource Availability for the Bioenergy Sector

The following section discusses the current UK context, barriers and potential strategies for increasing the availability of the UK's resources in the context of the research findings.

Strategies for Increasing the Availability of UK Resources Grown for the Bioenergy Sector

Research Outputs

The research identifies 'Crop and Agricultural Productivity' and the area of 'Land Dedicated for Energy Crop Growth' as the drivers that most significantly influence the availability of UK grown biomass resources such as energy crops.

The influence of realising higher limits of crop and agricultural productivity is shown to potentially increase the availability of this resource by >30% by 2050. This is an unsurprising trend, as greater crop yields will also benefit the production of crops dedicated for the energy sector. Although the standout driver with key influence on this resource is utilisation of available land dedicated for growth. Realising maximum levels of available land utilisation demonstrates a potential >87% improvement in resource availability in 2050, compared to conditions with reduced land use. Highlighting that if the UK wants to increase its biomass and energy crop resource, focussing on anything other than increasing land availability is unlikely to deliver the same scale of results.

Current UK Policy & Strategy Context

The UK Bioenergy Strategy [87] states that the increased growth of resources on unused or low ecosystem value lands is essential for producing resources for the bioenergy sector. Although the area of available land dedicated to grow these resources is essentially reliant on UK farmers utilising their lands to grow crops for the energy sector. To promote this the UK's primary incentive mechanism to promote farmers to grow biomass and energy crops has been the 'Energy Crops Scheme'. Although over the lifetime of the scheme widespread dedication of lands to grow biomass and energy crops has not materialised [94]. A summary of key barriers preventing land owner from producing resources for the bioenergy sector are presented in Table 5.

Pathways for Increasing Resource Availability

The demand for biomass and energy crops is growing fast [95], whilst their production offers environmental and economic benefits much wider than for just the energy sector [96]. Thus developing a policy framework and financial packages especially with respect to the Renewable Heat Incentive, Feed-in-Tariffs and a reworked Energy Crop Scheme are essential to reduce barriers and allow markets to drive progress [91].

The UK's already has good comparative examples of policies and incentives in the form of the Forestry Commission's 'Woodfuel Strategy' [97], where a roadmap and framework of targets backed by incentives are increasing the availability and use of woodfuels. There are also many examples of leading incentive schemes currently being applied across the EU to promote the bioenergy sector and incentivise the growth of resources [98]. These provide insights into further potential directions that the Government could go in developing UK policies.

Strategies for Increasing the Utilization of Agricultural Residues by the Bioenergy Sector

Research Outputs

'Population Change' and the 'Utilisation of Agricultural Residues' are the two drivers identified by the research as providing the greatest influencing the availability of this resource. These linkages appear to be self-evident, higher levels of population growth means that more food will need to be produced, resulting in greater availability of agricultural residues. At the same time the greater extent that agricultural residues are collected/harvested, the greater availability for the bioenergy sector.

However the more valuable analysis highlighted by Fig. 2 and also reflected within Fig. 4, is the near-term and continuous availability of agricultural residues – shown to be relatively constant and robust to major fluctuations caused by supply chain influences. The resource availability in 2015 is also forecast to exceed 10.3 Mt and steadily increase by >109% by 2050. Based on this analysis, agricultural residues should be highlighted and targeted within bioenergy strategies as reliable and robust opportunities for the bioenergy sector.

Current UK Policy & Strategy Policy Context

There are currently comparatively low levels of agricultural residues utilisation by the bioenergy sector in the UK. This trend is reflected in UK farming statistics [99] documenting that: <5% of the UK's livestock focused farms generate renewable energy, and of these <50% utilise manures and slurry feedstocks. Whilst <6% of arable agriculture focused farms generate renewable energy, and of these <45% utilise feedstocks such as straws. The UK Bioenergy Strategy [87] recognises the need to work to improve the economics of respective supply chains and bioenergy pathways, although many barriers remain as summarised in Table 5.

Potential Mechanisms for Increasing Resource Availability

There are many case studies that the UK could consider where agricultural residues are widely utilised by the bioenergy sector. Within Europe, Denmark represents the leading example of straw residue utilisation. Denmark's established harvesting infrastructure and market development is the consequence of targeted policy driven initiatives, such as: mandates requiring that higher prices are paid for energy from straws; collaborations between the bioenergy sector, individual farmer and specialised contractors enabling the shared utilisation of high specification harvesting and processing equipment; and a market structure that provides farmers with enhanced controls over their pricing demands, and standard contracts between produces and generators regardless of resource scale [100].

The utilisation of slurries and manures within anaerobic digestion bioenergy systems from large scale farms or localised farming cooperatives, represents key opportunities for the UK bioenergy sector. Raising awareness [101] and financial support [102] for these systems is key. The UK has an array of existing financial mechanisms and incentives [103] and [104] designed to promote this sector, although it remains highly undeveloped [105]. Again the UK could learn from successful policy case studies from across the EU, such as the German Renewable Energy Act [102] and related policies [105] and [106] that reduce the financial barriers of AD development schemes through directing increased financial responsibilities onto grid operators.

The European Common Agricultural Policy (CAP) is also widely identified as potential mechanism for increasing the utilisation of agricultural residues by the bioenergy sector. Potential reform areas being: further guidance of the quantities of resources to be returned to soils; initiatives to support residue supply chains; and the broadening of existing institutional and local partnerships to support the bioenergy sector[107].

Strategies for Increasing the Utilization of Household Waste Resources by the Bioenergy Sector

Research Outputs

The research finds that 'Waste Generation Trends' and 'Waste Management Strategies' are the key drivers determining the availability of household wastes. The analysis highlights that implementation of a waste management strategy that focuses on energy from waste pathways could provide over 40 Mt of household waste resource for the bioenergy sector by 2050. Household wastes therefore representing a substantial opportunity for the bioenergy sector, albeit highly reliant on the development of complementary waste management strategies.

Current UK Policy & Strategy Policy Context

Energy from waste in the UK has historically had a poor image with landfill distribution and early incinerators favoured. However the introduction of landfill diversion targets and the development of new technologies have placed energy from waste back on the UK's agenda. Although the prime focus of the UK's waste management strategies is to reduce and recycle, efficient energy recovery remains an important element of the strategy to both generate energy and reduce land-filled waste volumes [108]. A summary of key barriers preventing the wider utilisation of wastes and growth of the sector are presented in Table 5.

Table 5: Key barriers to the greater production & utilisation of UK resources by the bioenergy sector

Biomass resources	Barriers to increasing resource availability for the bioenergy sector
Grown Biomass & Energy Crops [91], [94] and [112]	Educational – awareness to incentive schemes, reluctance to move away from producing traditional agricultural crops, and poor understanding of energy crop establishment and management best practice.
	Economic – cash flow problems between planting and harvests, current margins associated with small scale productions, and the lack of links between biomass producers and markets.
	Legislative – lack of recognition of certain 'innovative' crops by inventive schemes.
	Technical – specific fuel requirements of bioenergy systems place increased demands on resources produced, and lack of processing infrastructure that would increase the economic viabilities.
Plant Based Agricultural Residues (straws) [100]	• Underdeveloped Markets – the lack of established supply chains for straw for bioenergy purposes.
	Competing Uses – straws are extensively used by existing markets with which the bioenergy sector will compete for resource.
	Inaccurate Guidance – overuse of the resource beyond best practice to maintain soil health can lead to large unnecessary impacts on resource availability.
	Undeveloped Infrastructure – the inaccessibility and lack of appropriate machinery and infrastructure for the handling and processing of straw residues.
	Resource Variability – due to varying climatic conditions and fluctuating harvest yields, the variability in the quantity and quality of straws has large implications for the bioenergy sector that typically requires specific fuel specifications.

Animal Based Agricultural Residues (slurries & manures) [101] and [113]	Transportation – The nature and bioenergy characteristics of slurries and manures makes them impractical, uneconomical and energy inefficient to be transported any great distance.
	Resource Availability – as a result of UK farming practices, slurry and manure can only typically by collected for a limited number of months, reflecting livestock housing regimes.
	Spatial Constraint – anaerobic digestion (AD) systems, the most suitable bioenergy systems for the use of manure and slurry resources require physical space. The economics of AD systems are also largely improved through the addition of energy crops feedstocks, which may require additional (potentially large) planting areas that are typically incompatible with the nature of the farms with the large animal based biomass resources.
	Capital Costs – the capital costs of digesters and associated infrastructure is high and are unlikely to fall significantly in the near-term.
	Collaboration Complexity – The time and costs associated with developing large community or district systems that pool resources from a number of local sites can be highly complex and costly.
Waste Resources [112], [114] and [115]	Incentive – the cost comparison of energy from waste systems compared to landfill represents a strong barrier against the further development of this sector.
	Waste Hierarchy – the supply of the specific waste feedstocks required by bioenergy systems is restricted by the waste hierarchy and the UK's waste policies primary focus to reduce and recycle.
	Opposition – social opposition led by local communities and the lobbying of environmental action groups are by far the greatest barrier to the development of the UK energy from waste sector.
	Finances – the varying definitions of biomass wastes and their respective subsidy regimes can prevent developers from accessing the finances required to grow the sector.

Potential Mechanisms for Increasing Resource Availability

The UK's scope for developing waste management strategies is highly restricted and defined by EU Directives [109]. However when adapting applicable EU Directives into national laws there is room for manoeuvre, with the definitions of wastes in the context of bioenergy being a key variable differing between Member States. Adjusting these categorisation parameters allows varying subsidisation and favourability of energy from waste pathways [109]. Reviewing these key policy variances between Member States presents a series of case studies for the UK to potentially consider if aiming to support the energy from waste sector.

With respect to addressing the large barriers associated with the social opposition to energy from waste technologies, the UK could draw influence from scenarios around the world and specifically other EU Member States, where public opinions are far less hostile. A review undertaken by WMW (2014) [110] found that: educating local populations of benefits, linking arguments to climate change, reassuring communities of air pollutant regulations and providing direct local energy benefits such as cheap district heat can vastly soften opposition. Also being mindful in planning processes that the voices of minority groups opposing energy from waste plant, often overshadow the opinions of the majority [111].

CONCLUSIONS

A Biomass Resource Model (BRM) was developed reflecting the UK's indigenous biomass supply chains. The drivers controlling the BRM were calibrated to 2050 to analyse current and forecast parameters in reflection of a wide literature review. The analysis focused on the development of a baseline scenario to determine the specific indigenous biomass resources that demonstrate the greatest potential for the UK bioenergy sector. Systematic analysis of the BRM's drivers allowed the evaluation of the extent that they influence indigenous resource availability to 2050.

Key policy conclusions for increasing the availability of UK indigenous resource for the bioenergy sector are highlighted below.

- Biomass and Energy Crops, Agricultural Residues and Household Wastes – are identified as the biomass resources that demonstrate the greatest promise for the UK bioenergy sector, in terms of their availability quantity and bioenergy potential.

- Potential and Mobilisation of Grown Biomass Resource – UK grown biomass and energy crop resources have been identified as potentially providing >31 Mt for the bioenergy sector by 2050. The standout driver influencing the availability of these resources was identified as the uptake of available land dedicated for its growth. However the analysis also highlighted that this resource currently has a relatively low starting base, with >1.9 Mt forecast by 2015. Therefore concerted efforts will be required in managing the drivers that influence availability, if anywhere near the upper levels of resource forecasts are to be realised. These should include the implementation of policies that encourage/ incentivise the utilisation of available land for the growth of resource dedicated for the bioenergy sector.

- Potential and Mobilisation of Biomass Residue Resource – Residue biomass resources were identified as potentially providing upto >29.8 Mt of resource for the bioenergy sector by 2050. Agricultural residues (straws & slurries) make up the majority of this quantity, whilst also continuing to be utilised to maintain soil systems. The availability of residues was forecast to steadily increase and be comparatively robust to supply chain influences. Biomass residues therefore representing a potentially continuous and reliable near and long-term indigenous resource option for the bioenergy sector.

- Potential and Mobilisation of Biomass Waste Resource – Waste biomass in the UK was identified as potentially providing up to >89 Mt of resource for the bioenergy sector by 2050. Household wastes being the largest waste contributor. Wastes were found to be highly influenced by one key driver, the waste management system adopted. The availability of waste resources was found to be much diminished when the adopted waste management strategy was uncomplimentary to the bioenergy sector. Therefore if wastes are to be increasingly utilised by the bioenergy sector, the analysis confirms the importance of implementing policies for effective development of waste management strategies.

- Refocusing Bioenergy Strategies to Increase the Availability of Indigenous Resources – The paper highlights the importance of applying a targeted approach for increasing the potential of indigenous resources. This is contrary to the broad policy focus approach currently being implemented in the UK. The analysis has identified that there are multiple biomass resource opportunities in the UK, but realisation of the upper levels of resource availability forecasts is highly dependent on the implementation of effective policies that target and manage the specific supply chain drivers most influential for each respective biomass resources.

APPENDIX A. SUPPLEMENTARY DATA

Supplementary data related to this article:

Appendix A1: Analysis of Biomass Resource Availability & Bioenergy Potential (Based on Literature Informed Forecasts)

	Resource Availability for Bioenergy Sector (Mt)				Bioenergy Generation Potential (TWh)			
	2015	2020	2030	2050	2015	2020	2030	2050
Biomass & Energy Crops	1.59	3.18	4.30	31.30	4.0	9.6	31.2	104.7
Dedicated Forestry Resources	1.09	1.82	4.30	1.98	1.9	3.5	9.3	5.0
Agricultural Residues	10.68	12.06	17.86	26.29	27.0	31.7	51.2	83.4
Arboricultural Residues	0.13	0.28	0.29	0.35	0.3	0.7	0.8	1.1
Forestry Residues	0.28	1.01	2.55	3.54	0.6	2.5	6.9	10.9
Industry Residues	0.59	0.60	0.67	0.81	1.5	1.6	2.0	2.7
Household Wastes	6.78	16.15	34.85	40.72	13.8	36.2	86.9	117.3
Food & Organic Wastes	2.63	5.38	11.19	14.70	1.3	2.9	6.8	10.3
Other Wastes	4.17	10.71	24.62	32.71	8.1	20.9	50.7	75.1
Sewage Wastes	1.66	1.73	1.88	2.12	2.8	3.3	3.9	5.1

Appendix A2: Analysis of the Extent that Drivers Influence the Availability of Biomass Resources

	Biomass & Energy Crops (Max Forecast Mt)				Agricultural Residues (Max Forecast Mt)				Household Wastes (Max Forecast Mt)			
	2015	2020	2030	2050	2015	2020	2030	2050	2015	2020	2030	2050
Population Change	0.86	1.76	5.89	18.30	10.31	11.22	16.54	24.11	2.74	3.86	6.04	6.79
Changes in Built-Up Land Area	0.85	1.69	5.53	16.69	10.22	10.99	15.76	21.42	2.74	3.86	6.04	6.79
Crop & Agriculture Productivity	1.02	2.11	7.14	21.82	10.22	10.99	15.77	21.43	2.74	3.86	6.04	6.79
Food Waste Generation	0.85	1.70	5.54	16.71	10.22	10.99	15.76	21.42	2.74	3.86	6.04	6.79
Food Commodity Imports	0.85	1.69	5.52	16.66	10.22	10.99	15.76	21.42	2.74	3.86	6.04	6.79
Food Commodity Exports	0.85	1.69	5.52	16.66	10.22	10.99	15.76	21.42	2.74	3.86	6.04	6.79
Utilisation of Agricultural Wastes & Residues	0.85	1.69	5.52	16.66	10.60	11.85	17.15	23.89	2.74	3.86	6.04	6.79
Forestry Expansion & Productivity	0.86	1.73	5.61	16.66	10.22	10.99	15.76	21.42	2.74	3.86	6.04	6.79
Wood-based Industry Productivity	0.85	1.69	5.52	16.66	10.22	10.99	15.76	21.42	2.74	3.86	6.04	6.79
Imports of Forestry Product	0.85	1.69	5.52	16.66	10.22	10.99	15.76	21.42	2.74	3.86	6.04	6.79
Exports of Forestry Product	0.85	1.69	5.52	16.66	10.22	10.99	15.76	21.42	2.74	3.86	6.04	6.79
Utilisation of Forestry Residues	0.85	1.69	5.52	16.66	10.22	10.99	15.76	21.42	2.74	3.86	6.04	6.79
Utilisation of Industrial Residues	0.85	1.69	5.52	16.66	10.22	10.99	15.76	21.42	2.74	3.86	6.04	6.79
Utilisation of Arboriculture Arisings	0.85	1.69	5.52	16.66	10.22	10.99	15.76	21.42	2.74	3.86	6.04	6.79
Waste Generation Trends	0.85	1.69	5.52	16.66	10.22	10.99	15.76	21.42	2.79	4.00	6.77	8.51
Waste Management Strategies.	0.85	1.69	5.52	16.66	10.22	10.99	15.76	21.42	6.78	16.15	34.85	40.72
Land Dedicated for Energy Crop Growth	1.59	3.18	10.37	31.30	10.23	10.99	15.79	21.51	2.74	3.86	6.04	6.79

REFERENCES

1.   Deloitte. Knock on wood: is biomass the answer to 2020?. London, United Kingdom: Deloitte LPP; 2012. p. 1e24.

2.   Panoutsou C, Eleftheriadis J, Nikolaou A. Biomass supply in EU27 from 2010 to 2030. Energy Policy 2009; 37(12):5675e86.

3.   Directive 2009/28/EC On the promotion of the use of energy from renewable sources and amending and subsequently repealing directives 2001/77/EC and 2003/30/EC. 2009.04.23. Off J Eur Union 2009; L 140:16e62.

4.   DECC. The UK renewable energy strategy. London, United Kingdom: Department for Energy and Climate Change; 15.07.2009.

5.   EUROSTAT. Economy-wide material flow accounts and derived indicators: a methodology guide. Luxembourg: European Communities; 2011.

6.   Upham P, Riesch H, Tomei J, Thornley P. The sustainability of forestry biomass supply for EU bioenergy: a post-normal approach to environmental risk and uncertainty. Environ Sci Policy 2011;14(6):510e8.

7.   Hewitt J. Flows of biomass to and from the EU: an analysis of data and trends. Brussels, Belgium: FERN; 2011 July. p. 1e54.

8.   RSPB. Bioenergy: a burning issue. Scotland: Royal Society for the Protection of Birds; 2011. p. 1e22.

9.   Welfle A, Gilbert P, Thornley P. Securing a bioenergy future without imports. Energy Policy 2014;68:1e14.

10.   IEEP. Securing biomass for energy e developing an environmentally responsible industry for the UK now and into the future. London, United Kingdom: Institute for European Environmental Policy; 2011. Sponsored by RSPB, Friends of the Earth, Greenpeace UK, Woodland Trust.

11.   UK DEA. Waste wood, the untapped resource for biomass fuel. Cirencester, United Kingdom: UK District Energy Association; 2008.

12.   Greenpeace. Decentralising UK energy: cleaner, cheaper, more secure energy for 21st century Britain. London, United Kingdom: Greenpeace; 2006.

13.   SDC. Wood fuel for warmth. Scotland: Sustainable Development Commission; 2005.

14.   Biomass Energy Centre. Resource availability. Surrey, United Kingdom: Biomass Energy Centre; 2013.

15.   AEA. UK and global bioenergy resource e final report. London, United Kingdom: AEA; 2011. Sponsored by Oxford Economics; Biomass Energy Centre; Forest Research; Department for Energy & Climate Change.

16.   AEA. UK and global bioenergy resource e annex 1 report: details of analysis. London, United Kingdom: AEA; 2010. Sponsored by Oxford Economics; Biomass Energy Centre; Forest Research; Department for Energy & Climate Change.

17.   Welfle A, Gilbert P, Thornley P. In: Meeting bioenergy targets with reduced imports: proceedings of the 21st European biomass conference and exhibition; 2013 June 03-07; Copenhagen, Denmark.

18.   Thran D, Seidenberger T, Zeddies J, Offermann R. Global biomass potentials e resources, drivers & scenario results. Energy Sustain Dev 2010;14(3):200e5.

19.   Bottcher H, Frank S, Havlik P. Biomass availability & supply analysis. Laxenburg, Austria: IIASA; 2012. Biomass Futures Draft Deliverable 3.4, IEE08653SI2.529241. Funded by the EU's Intelligent Energy Programme.

20.   Long H, Li X, Wang H, Jia J. Biomass resources & their bioenergy potential estimation: a review. Renew Sustain Energy Rev 2013;26:344e52.

21.   Slade R, Saunders R, Gross R, Bauen A. Energy from biomass: the size of the global resource. London, United Kingdom: UK Energy Research Centre; 2011.

22.   Fischer G, Schrattenholzer L. Global bioenergy potentials through 2050. Biomass Bioenergy 2001;20(3):151e9.

23.   Haberl H, Beringer T, Bhattacharya S, Erb K, Hoogwijk M. The global technical potential of bio-energy in 2050 considering sustainability constraints. Curr Opin Environ Sustain 2010;2(5e6):394e403.

24.   Hoogwijk M, Graud W. On the global & regional potential of renewable energy sources. London, United Kingdom: Ecofys UK;

2004 March. Sponsored by REN21-Renewable Energy Policy Network for the 21st Century.

25. Lysen E, van Egmond S, Dornburg V, Faaij A, Verweij P, Langeveld H, et al. Biomass assessment: assessment of global biomass potentials & their links to food, water, biodiversity, energy demand & economy. Netherland: Universiteit Utrecht, Wageningen UR, NEAA, Universiteit Amsterdam, ECN, UCE; 2008 January. Sponsored by the Netherlands Research Programme on Scientific Assessment and Policy Analysis for Climate Change.

26. Smeets E, Faaij A. Bioenergy potentials from forestry in 2050: an assessment of the drivers that determine the potentials. Clim Change 2007;81:353e90.

27. Adams P, Hammond G, McManus M, Mezzullo W. Barriers to & drivers for UK bioenergy development. Renew Sustain Energy Rev 2011;15:1217e27.

28. BR&Di. Increasing feedstock production for biofuels: economic drivers, environmental implications, & the role of research. Washington, United States: Biomass Research & Development Initiative; 2011.

29. Ladanai S, Vinterback J. Global potential of sustainable biomass for energy. Uppsala, Sweden: Department of Energy and Technology, Swedish University of Agricultural Sciences; 2009.

30. Haberl H, Erb K, Krausmann F, Bondeau A, Lauk C, Muller C, et al. Global bioenergy potentials from agricultural land in 2050: sensitivity to climate change, diets and yields. Biomass Bioenergy 2011;35(12):4753e69. biomass and bioenergy 70 (2014) 249 e266 263

31. Bruinsma J. World agriculture: towards 2015e2030. Rome, Italy: Food and Agriculture Organisation; 2003. FAO Perspective Report.

32. United Nations. World population prospects. New York, United States: United Nations Populations Division; 2010.

33. Muller A, Schmidhuber J, Hoogeveen J, Steduto P. Some insights in the effect of growing bio-energy demand on global food security and natural resources. Rome, Italy: Food and Agriculture Organisation; 2007.

34. Prieler S. Built-up and associated land area increases in Europe. Laxenburg, Austria: International Institute for Applied Systems Analysis; 2011. MOSUS Model: WP3 e Environmental Evaluation.

35. Taylor G. Bioenergy for heat and electricity in the UK: a research atlas and roadmap. Energy Policy 2008;36(12):4383e9.

36. Mueller N, Gerber J, Johnston M, Ray D, Ramankutty N, Foley J. Closing yield gaps through nutrient and water management. Nature 2012; 490:254e7.

37. Soil Association. Telling porkies e the big fat lie about doubling food production. Bristol, United Kingdom: Soil Association; 2010. p. 1e11.

38. Roder M, Thornley P, Campbell G, Bows-Larkin A. Emissions associated with meeting the future global wheat demand: a case study of UK production under climate change constraints. Environ Sci Policy 2014; 39:13e24.

39. FAOSTAT. FAO statistics food commodity datasets. Food and Agriculture Organisation; 2011. Available from: http:// faostat. fao.org. Files Updated Annually.

40. EUROSTAT. EUROSTAT e forestry statistics database. European Commission Statistics Database; 2012. Available from: http:// ec.europa.eu/eurostat;. Files Updated Annually.

41. DEFRA. Farming statistics: agriculture in the United Kingdom. London, United Kingdom: Department for Environment, Food and Rural Affairs; 2012 May.

42. Bouwman A, Van der Hoek W. Exploring changes in world ruminant production systems. Agric Syst 2004; 84(2):121e53.

43. Smeets E, Faaij A, Lewandowski I. A quickscan of global bioenergy potentials to 2050. Utrecht, Netherlands: Copernicus Institute of Sustainable Development, Universiteit Utrecht; 2004.

44. Thornley P, Tomei J, Upham P. Theme 6 resource assessment feedstock properties. Manchester, United Kingdom: Supergen Biomass and Bioenergy Consortium, University of Manchester; 2008.

45. C-Tech Innovation. United Kingdom food and drink processing mass balance. Loughborough, United Kingdom: Biffaward Programme on Sustainable Resource Use; 2004.

46. Green A, Johnston N. Food surplus; reduction, recovery and recycle. Norwich, United Kingdom: Total Foods; 2004.

47. Nellman C, MacDevette M, Manders T, Eickhout B, Svihus B, Prins A. The environment food crisis e the environment's role in averting future food crises. Arendal, Norway; Cambridge, United Kingdom: United Nations Environment Programme, UNEP/GRID-Arendal, UNEP-WCMC; 2009.

48. DEFRA. Government review of waste policy in England 2011. London, United Kingdom: Department for Environment, Food and Rural Affairs; 2011 June. Ref: PB13540.

49. European Commission. Reducing the amount of food wasted by European consumers each year. Brussels, Belgium: European Commission; 2012.

50. Foresight. Foresight project on global food and farming futures e synthesis report C4: food system scenarios and modelling. London, United Kingdom: UK Government Office for Science; 2011.

51. Foresight. The future of food and farming: challenges and choices for global sustainability. London, United Kingdom: UK Government Office for Science; 2011.

52. Godfray H, Beddington J, Crute I, Haddad L, Lawrence L, Muir J, et al. Food security: the challenge of feeding 9 billion people. Science 2010;327:812e8.

53. WRI. Reducing food loss and waste. Washington, United States: World Resources Institute, Washington; 2013.

54. House of Commons Environment Food and Rural Affairs Committee. Securing food supplies up to 2050: the challenges faced by the UK. London, United Kingdom: House of Commons; 2009 July. Fourth Report of Session 2008/9: 1.

55. DEFRA. Driving export growth in the farming, food and drink sector: a plan of action 2012. London, United Kingdom: Department for Environment, Food and Rural Affairs; 2012.

56. BIS. Industrial strategy: UK sector analysis. London, United Kingdom: Department for Business Innovation & Skills; 2012. Economics Paper No.18.

57. CSL. National and regional supply/demand balance for agricultural straw in Great Britain. York, United Kingdom: Central Science Laboratory; 2008.

58. E4tech. Biomass supply curves for the UK. London, United Kingdom: E4tech; 2009. Sponsored by the Department for Energy and Climate Change.

59. DEFRA. UK biomass strategy. London, United Kingdom: Department of Trade and Industry, Department for Transport, Department for Environment Food and Rural Affairs; 2007 May..

60. DEFRA. Farm practices survey. London, United Kingdom: Department for Environment, Food and Rural Affairs; 2011 August.

61. DEFRA. The economics of waste and waste policy. London, United Kingdom: Environment and Growth Economics, Department for Environment, Food and Rural Affairs; 2011 June. Ref PB13548.

62. Smith K, Charles D, Moorhouse D. Nitrogen excretion by farm livestock with respect to land spreading requirements and controlling nitrogen losses to ground surface waters e part 1: cattle and sheep. Bioresour Technol 2000;71:173e81.

63. Smith K, Charles D, Moorhouse D. Nitrogen excretion by farm livestock with respect to land spreading requirements and controlling nitrogen losses to ground surface waters e part 2: pigs and poultry. Bioresour Technol 2000;71:183e94.

64. Kilpatrick J. Addressing the land use issues for non-food crops, in response to increasing fuel and energy generation opportunities. Hereford, United Kingdom: ADAS; 2008. NNFCC Project 08-004. Sponsored by Defra; Managed by NNFCC.

65. Fischer G, Hizsnyik E, Prieler S, Van Velthuizen H. Assessment of biomass potentials for biofuel feedstock production in Europe: methodology and results. European Union Refuel Program; 2007.

66. DECC. Bioenergy strategy analytical annex. London, United Kingdom: Department for Energy and Climate Change; 2012.

67. Forestry Commission. Forestry statistics: a compendium of statistics about woodland, forestry and primary wood processing in the UK. Bristol, United Kingdom: Forestry Commission; 2012.

68.   Jenkins T, Matthews, Mackie E, Halsall L. Restocking in the forecast. Bristol, United Kingdom: Forestry Commission; 2012. Forecast Technical Document PF2011.

69.   National Forest Inventory, Forestry Commission. GB 25 year forecast of standing coniferous volume and increment. Edinburgh, United Kingdom: Forestry Commission, National Forest Inventory; 2012 July. National Forest Inventory Report.

70.   Jenkins T, Matthews, Mackie E, Halsall L. Forecast types. Bristol, United Kingdom: Forestry Commission; 2012. Forecast Technical Document PF2011. 264 biomass and bioenergy 70 (2014) 249 e266

71.   Jenkins T, Matthews, Mackie E, Halsall L. Growing stock volume forecasts. Bristol, United Kingdom: Forestry Commission; 2012. Forecast Technical Document PF2011.

72.   Forestry Commission. 25 Year forecast of softwood timber availability. Bristol, United Kingdom: Forestry Commission; 2012. National Forest Inventory Report.

73.   Forestry Commission. Volume increment forecasts. Bristol, United Kingdom: Forestry Commission; 2012. Forecast Technical Document.

74.   Jenkins T, Matthews, Mackie E, Halsall L. Felling and removals forecasts. Bristol, United Kingdom: Forestry Commission; 2012. Forecast Technical Document PF2011.

75.   WPIF. Make wood work. Grantham, United Kingdom: Wood Panel Industries Federation; 2012.

76.   WPIF. Wood fibre availability and demand in Britain 2007 to 2025. Grantham, United Kingdom: Wood Panel Industries Federation; 2010.

77.   McKay H. Woodfuel Resource in Britain. Forestry Contracting Association; 2003. Sponsored by Department for Trade and Industry, Scottish Enterprise, Welsh Assemble, Forestry Commission. FES B/W3/00787/REP/1, DTI/Pub URN 03/1436.

78.   Stupak I, Lattimore B, Titus B, Smith T. Criteria and indicators for sustainable forest fuel production and harvesting: a review of current standards for sustainable forest management. Biomass Bioenergy 2011;35(8):3287e308.

79.   Lattimore B, Smith C, Titus B, Stupak I, Egnell G. Environmental factors in woodfuel production: opportunities, risks, and criteria and indicators for sustainable practices. Biomass Bioenergy 2009; 33(10):1321e42.

80.   Perlack RD, Wright LL, Turhollow AF, Graham RL, Stokes BJ, Erbach DC. Biomass for a feedstock for bioenergy and bioproducts industry: the technical feasibility of a billion ton annual supply. Oak Ridge, Tennessee: Oak Ridge National Laboratory; 2005 May. DOE/GO-102995-2135, ORNL/TM-2005/66.

81.   Jablonski S, Strachan N, Brand C, Pantaleo A, Bauen A. A systematic assessment of bioenergy representation in the Markal model: insights on the formulation of bioenergy scenarios. London; Oxford, United Kingdom: Imperial College London, Kings College London, Oxford Environmental Change Institute; 2008.

82.   Forestry Commission. Wood fuel resources. Bristol, United Kingdom: Forestry Commission; 2013.

83.   WRAP. Wood waste market in the UK e summary report. Banbury, United Kingdom: Waste and Resources Action Programme; 2009 August. Project code: MKN022.

84.   WRAP. Wood waste market situation report. Banbury, United Kingdom: Waste and Resources Action Programme; 2011 April.

85.   ERM, Golder Associates. Carbon balances and energy impacts of the management of UK waste. London, United Kingdom: Environmental Resources Management; Golder Associates; 2006. Sponsored by Department for Environment, Food and Rural Affairs. R&D Project WRT 237.

86.   DEFRA. Consultation on recovery and recycling targets for packaging waste for 2013-2017 e a consultation on proposed changes to the producer responsibility obligations (packaging waste) regulations 2007 (as amended) and the producer responsibility obligations (packaging waste) regulations (Northern Ireland) 2007 (as amended). London, United Kingdom: Department of the Environment, The Scottish Government, Welsh Government, Department for Environment Food and Rural Affairs; 2011.

87.   DECC. UK bioenergy strategy. London, United Kingdom: Department of Energy and Climate Change, Department for

Environment Food and Rural Affairs, Department for Transport; 2012 April.

88. Rowe R, Whitaker J, Chapman J, Howard D, Taylor G. Life cycle assessment in the bioenergy sector: developing a systematic review. London, United Kingdom: UK Energy Research Centre; 2008:01:21. UKERC/WP/FSE/2008/002.

89. Rowe R, Street N, Taylor G. Identifying potential environmental impacts of large-scale deployment of dedicated bioenergy crops in the UK. Renew Sustain Energy Rev 2009; 13(1):271e90.

90. Biomass Task Force. Biomass task force. London, United Kingdom: Task Force Report to Government; 2005 October. Sponsored by the Department for Environment, Food and Rural Affairs.

91. NNFCC. Domestic energy crops: potential and constraints review. York, United Kingdom: National Non-Food Crops Centre; 2012. Project Number: 12-021, URN: 12D/081. Sponsored by the Department for Energy & Climate Change.

92. EEA. Estimating the environmentally compatible bioenergy potential from agriculture. Copenhagen, Denmark: European Environment Agency; 2007. Technical Report No. 12/2007.

93. ADAS. Addressing the land use issues for non-food crops, in response to increasing fuel and energy generation opportunities. ADAS; 2008. NNFCC Project 08-004.

94. Natural England. Energy crop scheme. London, United Kingdom: Natural England; 2014. Accessed online at: www. naturalengland. org.uk/ourwork/farming/funding/ecs/.

95. DEFRA. Rural development programme for England. London, United Kingdom: Department for Environment, Food & Rural Affairs; 2013.

96. Forestry Commission. Energy crops e benefits of green space. Bristol, United Kingdom: Forestry Commission; 2013.

97. Forestry Commission. A woodfuel strategy for England. Bristol, United Kingdom: Forestry Commission; 2013.

98. Jyvaskyl a Innovation Oy, MTT Agrifood Research. Energy from field energy crops e a handbook for energy producers. Jyvaskyl a, Finland: Jyv askyl a Innovation Oy, MTT Agrifood Research; Finland 2009. Sponsored by Intelligent Energy, European Communities.

99. DEFRA. Farming statistics e diversification & renewable energy production on farms in England 2010. London, United Kingdom: Department for Environment, Food & Rural Affairs; 2013.

100. Kretschmer B, Allen B, Hart K. Mobilising cereal straw in the EU to feed advanced biofuel production. London, United Kingdom: Institute for European Environmental Policy; 2012 May. Sponsored by Novozymes.

101. Bywater A. A review of anaerobic digestion plants on UK farms e barriers, benefits & case studies. Kenilworth, United Kingdom: Royal Agricultural Society of England; 2012.

102. The German Government. Erneuerbare-Energien-Gesetz EEG. Berlin, Germany: Federal Ministry of Economics and Technology; 2000.

103. DEFRA & DECC. Anaerobic digestion strategy & action plan. London, United Kingdom: Department for Environment, Food & Rural Affairs; Department for Energy & Climate Change; 2011.

104. DEFRA. Anaerobic digestion strategy & action plan e annual report 2012-13. London, United Kingdom: Department for Environment, Food & Rural Affairs; 2013.

105. Humphries A. Implementation of anaerobic digestion in the UK, California & Germany. London, United Kingdom: Farmers Club; 2011 June.

106. Volk G. Biogas and the German Ordinance of Gas Network Access (GasNZV). Berlin, Germany: Bundesnetzagentur; 2011.

107. Kretschmer B, Buckwell A, Smith C, Watkins E, Allen B. Recycling agricultural, forestry & food wastes & residues for biomass and bioenergy 70 (2014) 249 e266 265sustainable bioenergy & biomaterials. London, United Kingdom: Institute for European Environmental Policy; 2013. Sponsored by the European Parliament.

108. DEFRA. Energy from waste. London, United Kingdom: Department for Environment, Food & Rural Affairs; 2014.

109. L. Fagernäs, A. Johansson, C. Wilén, K. Sipilä, T. Mäkinen, S. Helynen, et al.Bioenergy in Europe – opportunities & barriers VTT Technical Research Centre of Finland, Helsinki, Finland (2006).

110. WMW. The burning question: how is the public perception of waste to energy improving. Northbrook, United States: Waste Management World; 2014. Accessed online at: www. waste-management-world.com/articles/print/volume-12/ issue-4/ features/the-burning-question-how-is-the-publicperception-of-waste-to-energy-improving.html.

111. Bredee H, Georgeson R. Public attitudes to community buy in for waste & resource infrastructure. Maidenhead, United Kingdom: Ray Georgeson Resources Ltd; GfK NOP; 2011 June. Sponsored by SITA UK.

112. McDermott F, Hopwood L, Evans G. Barriers to bioenergy matrix. York, United Kingdom: National Non-Foods Crop Centre; 2012.

113. de Buisonje F. Manure management & processing. The Netherlands: Wageningen UR Livestock Research; 2013.

114. WMW. UK urged to supply more energy through waste. Northbrook, United States: Waste Management World; 2011. Accessed online at: www.waste-management-world. com/ articles/print/volume-11/issue-6/regulars/news/ukurged-to-supply-more-energy-through-waste.html.

115. Cheeseman C, Velis C. WTERT-UK opportunities & barriers. London, United Kingdom: Department of Civil and Environmental Engineering Imperial College; 2010.

Chapter 4

Harvesting Microalgae by Ctab-Aided Foam Flotation Increases Lipid Recovery And Improves Fatty Acid Methyl Ester Characteristics

Thea Coward[a], Jonathan G.M., Lee[a], and Gary S. Caldwell[b]

[a]School of Chemical Engineering and Advanced Materials, Merz Court, Newcastle University, Newcastle upon Tyne, NE1 7RU, England, UK

[b]School of Marine Science and Technology, Ridley Building, Newcastle University, Newcastle upon Tyne, NE1 7RU, England, UK

ABSTRACT

Foam flotation is an effective and energy efficient method of harvesting microalgae. This study has investigated the influence of growth phase and lipid content on harvesting efficiency. The highest biomass concentration factors were gained during active culture growth.

Surprisingly, the quantities of lipid recovered from microalgae harvested by foam flotation using the surfactant cetyl trimethylammonium bromide (CTAB), were significantly higher than from cells harvested by centrifugation. Further, cells harvested by CTAB-aided foam flotation exhibited a lipid profile more suited to biodiesel conversion containing increased levels of saturated and monounsaturated fatty acids. The enhanced lipid recovery was partially explained by the interaction of the cells with the surfactant, CTAB, which adsorbed onto the algae and was carried over into the total lipid extraction process. However, further evidence also suggested that CTAB promoted in situ cell lysis by solubilizing the phospholipid bilayer, thus increasing the amount of extractable lipid. This work demonstrates substantial added value of foam flotation as a microalgae harvesting method beyond energy efficient biomass recovery.

INTRODUCTION

Microalgae are considered as having great potential as a sustainable, scalable and affordable source of bio products including biofuels and high value chemicals [1], [2] and [3]. With the exception of microalgae grown heterotrophic ally [4], bulk products derived from algae, such as biofuels, are not yet cost competitive. Two major challenges facing the nascent microalgae industry are: 1) the ability to reliably and affordably produce high biomass yields at scale; and, 2) the harvesting, dewatering, and extraction of biochemical from large quantities of feedstock [2] and [5]. Concerning high volume, low value products i.e. algae biodiesel, improvements in production, harvesting, and integrating co-production of higher-value products/processes as part of a bio refinery concept will be critical to reduce algae oil production cost [3], [6],[7] and [8].

In recent years, considerable research efforts have focused on the application of harvesting technologies to improve biomass recovery; however, the majority of harvesting methods are not sufficiently cost effective for the economical production of low value outputs [9] and [10]. Flotation, a separation process originating from the mineral industry, has become an established method to remove algae from suspension [11], [12],[13], [14], [15] and [16], with potential for application to harvest microalgae for biofuel production [2], [5],

[6],[17], [18] and [19]. Flotation, particularly dissolved air flotation, is the favored technique for algae removal [5], [16], [20] and [21]. During dissolved air flotation an inorganic flocculent such as alum is added to aggregate the cells. Small bubbles formed by super saturation of the water with air, adhere to and transport suspended biomass to the surface [5], [13], [16], [19] and [22].

However, despite its effectiveness, dissolved air flotation is energy intensive, consuming up to 7.6 kWh m^{-3} [16]. In contrast, foam flotation, a technique similar to dissolved air flotation, virtually eliminates the need for the energy intensive step of air compression by generating bubbles and foam with the addition of a cationic surfactant e.g. cetyl trimethylammonium bromide (CTAB) and a low pressure sparer or agitator [16] and [19]; operational energy consumption is reduced to as little as 0.015 kWh m^{-3} [10]. CTAB has proven an advantageous surfactant for algae removal [10], [11], [13], [14], [15], [21] and [23] with reported removal efficiencies approaching 90% [11], [13], [14] and [15]. Additionally, the surfactant, by improving the electrostatic interactions between the bubbles and the microalgae cells, removes the need for a flocculent [21]. There are two main hypotheses as to how surfactants improve electrostatic interactions between bubbles and algae (Fig. 1).

The first predicts that cationic surfactants modify the surface properties of the bubble by establishing a positive charge, and therefore electrostatically attracted to the negatively charged algae [20], [21] and [24]. The second hypothesis predicts that cationic surfactant ions adsorb onto the algae making the cell hydrophobic and therefore available for bubble attachment [11], [13] and [15].

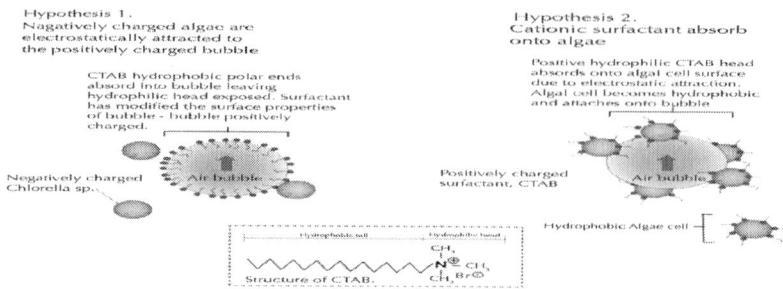

Figure. 1: The two proposed hypotheses as to how surfactants improve electrostatic interactions between bubbles and algae. In hypothesis 1, the cationic CTAB causes the bubble to become positively charged facilitating an electrostatic attraction to the negatively charged microalgae cells. In hypothesis 2, CTAB adsorbs onto the microalgae thereby making the cells hydrophobic and promoting bubble attachment (Diagram based on the works of Phoochinda et al. [15], and Zhang et al. [31]).

Despite the obvious appeal of flotation for harvesting microalgae, there remains limited evidence of its direct application in situations wherein the biomass is a valued product as opposed to a waste/nuisance. It is vital to determine the effects of harvesting on biomass quality and composition, particularly when the biochemical components must achieve defined quality standards for subsequent downstream processing [25], [26] and [27]. For example, Borges et al. [28] observed that by using a cationic flocculent to harvest the diatom Thalassiosira weissflogii higher percentages of C16:0, C16:1 and C20:5 fatty acids were recovered, whereas the yields of C18:0 and C18:1n9c decreased. This situation is further complicated by the fact that relatively few microalgae biochemical of commercial interest are produced constitutively; rather, the biosynthesis of many biochemical is induced during specific growth phases or in response to exogenous stimuli such as nitrogen deprivation [29]. To date, the influence of growth phase on the efficiency of settling, tangential flow filtration [30], flocculation [32], and the coagulant dosage required for dissolved air flotation has been investigated [18]. However, there is currently no information available on the possible effects of growth phase on the efficiency of CTAB-aided foam flotation. As indicated by Daquan et al. [30] and Zhang et al. [31] the significance of this cannot be discounted when considering the applicability and efficacy of foam flotation. Flotation is reliant upon the electrochemical properties of the algae cell;

however, the strength of this charge – measured as the zeta potential – is non-uniform, varying with cell age [31]. The optimal zeta potential may therefore not necessarily be in synchrony with the biosynthetic dynamics of the algae product(s) of interest, potentially resulting in a product-harvest mismatch thus reducing overall operational efficiency. The aims of this study were, therefore, to establish the influence of growth phase and CTAB exposure on foam flotation efficiency with respect to biomass and lipid recovery, and overall fatty acid profile.

MATERIALS AND METHODS

Cultivation of Algae

A non-axenic Chlorella sp. obtained from Blades Biological Ltd., Kent, UK, was grown in batch with an initial cell density of 10^7 cm^3 using a protease peptone medium containing the following (per liter): 0.2 g $MgSO_4 \cdot 7H_2O$, 0.2 g K_2HPO_4, 2 g KNO_3 and 1 g protease peptone (oxoid L85) (Sigma Aldrich). Cultures were grown in 20 L polycarbonate carboys (Nalgene) at 19 ± 2 °C with a 16 L: 8 D photoperiod. Lighting was supplied by a combination of warm and cold fluorescent tubes giving a luminance range of between 2200 and 2800 Lux. Mixing and gas transfer was facilitated by an aquarium pump. Growth was determined by cell counts using a haemocytometer every 3 days over a 21 day culture period.

Harvesting Experiments

Cells from the same culture were harvested at 3 day intervals between days 3 and 21 using two separate methods: foam flotation and centrifugation. The flotation column was previously described in Coward et al. [10]. During foam flotation 2 L of culture were mixed with 7.5 L of tap water to give an initial dry weight of 126.14 ± 10.06 mg L^{-1}, which is similar to dry weight yields from paddle–wheel open culture raceway systems [33]. All foam flotation harvests were conducted under the following conditions: column height of 0.5 m, air flow rate of 100 L h^{-1}, 30 min batch run time, and 10 mg L^{-1} of CTAB [10]. CTAB dissolved within 500 cm^3 of water was added to the harvest chamber along with the diluted algae to make an initial starting volume of 10 L. A ceramic flat plate sparger with a mean pore diameter of 20.0 μm [34] and dimensions of 2 cm × 11 cm × 11 cm in height, width, and length respectively, was used to generate bubbles with a mean diameter of 860 ± 158 μm. The bubble size was determined using a combination of a high-speed digital video camera (Olympus i-speed 3) and image processing software Image J (National Institutes of Health, Bethesda, Maryland, USA). Each harvest had 4 replicate runs. The concentration factor (CF) of each harvest was calculated as described in Coward et al. [10].

To prepare the cells for freeze drying and to remove any potential CTAB residue the foam flotation harvests were further concentrated by centrifugation (Sigma 3K18C) for 20 min at 8700× g. The cell pellets were washed in distilled water and centrifuged again. The washing step was repeated once more before the pellet was freeze dried for 48 h at 3 kPa in a Christ Alpha 1-4 LD Plus (SciQuip, UK), with a condenser temperature of −55 °C. To compare harvest methods 2 L of culture were harvested by centrifugation only (20 min at 8700× g) and then freeze dried, in 4 replicate runs.

To compare the effects of foam flotation and centrifugation on lipid recovery and fatty acid profiles, additional harvests were conducted from the same cultures after 12 days of growth, as this is when the highest concentration factor was gained (see Fig. 2).

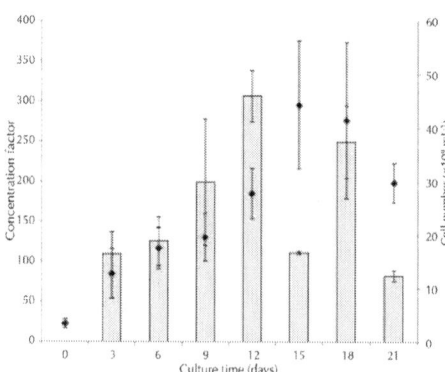

Figure. 2: Growth profile (diamonds) and concentration factors (light grey bars) for Chlorella sp. harvested by foam flotation across a 21 day period. Mean ± standard deviation.

Surfactant Adsorption

A methyl orange colorimetric test was conducted to assess the extent of CTAB adsorption onto the algae cells. An average culture dry weight of 350 ± 120 mg L^{-1} was gained by adding two hundred milliliters of Chlorella, which was over 21 days into the culture, to 750 cm^3, with three replicates. The volume of each replicate was made up to 1 L by adding 10 mg of CTAB dissolved in 50 cm^3 of deionized water. The Chlorella was exposed to the CTAB solution for 1 h during which the mixtures were stirred continuously using a magnetic stirrer. A 1 h exposure was determined as an approximation of the total duration the cells would be exposed to CTAB during the flotation harvesting process. After 1 h, 40 cm^3 (V_1) of each algae/surfactant solution was centrifuged for 20 min at 8700× g in replicates of four. For each replicate the supernatant was collected for analysis and the pellet re-suspended in a known volume of deionized water (V_2). This washing step was repeated for a third time (V_3). For analysis, 10 cm^3 of each supernatant sample was treated with 50 cm^3 of chloroform and an excess of methyl orange reagent under acidic conditions as described by Wang and Langley [35]. The methyl orange reacted with CTAB forming a chloroform soluble complex. The intensity of color produced in the chloroform layer was directly proportional to the concentration of methyl orange–surfactant complex concentration when measured spectrophotometrically at 415 nm. A calibration curve was created for CTAB in the 2.5–10 mg L^{-1} range, yielding an R^2 of 99.78% (data not shown).

The percentage of CTAB adsorbed onto the algae (ACTAB) was calculated using equation (1); where V_x is the volume of supernatant removed per centrifugal recovery, and MO_x is the quantity of CTAB present in the analyzed supernatant.

Equation (1)

$$A_{CTAB} = \left(\frac{V_1 MO_1 + V_2 MO_2 + V_3 MO_3}{V_1} \right) \times 100 \qquad (1)$$

Total Lipid Extraction

A modified version of the Folch method [36] was used for lipid extraction. The freeze dried microalgae was ground to a fine power using a pestle and mortar. A known quantity of microalgae, 0.095–0.33 g, was homogenized with methanol, followed by chloroform (1:2 v/v). The total homogenate volume was 30 times that of the tissue weight. The homogenates were centrifuged for 20 min at 4500× g to remove cell debris. The lipid fraction was transferred to a clean test tube, using a Pasteur pipette. A 0.88% potassium chloride solution was then added at 25% of the starting volume to wash the lipid fraction and remove any non-lipid contaminants. The final biphasic system was centrifuged for 20 min at 4500× g, and the resulting mixture left to separate into two phases. The lower phase was transferred into a pre-weighed glass tube and evaporated to dryness under a nitrogen stream at 37 °C. The weight of the crude lipid obtained from each sample was determined gravimetrically.

Solid Phase Extraction

The total lipid extracts were separated into lipid classes by solid phase extraction following the methods of Kaluzny et al. [37]. Total lipid mixtures in chloroform were applied to EASY® cartridges (Chroma bond 3 cm^3, 200 mg, Machete–Nagel, Germany), which had previously been conditioned with hexane. Solvent mixtures of increasing polarity were used to elute individual lipid classes. Neutral lipids were eluted with 4 cm^3 chloroform and propanol (2:1), free fatty acids with 4 cm^3 of 2% acetic acid in diethyl ether, and the phospholipid fractions with 4 cm^3 of methanol. Each lipid fraction was collected in pre-weighed receiving tubes and evaporated to dryness under nitrogen. The mass of each lipid class was determined gravimetrically [38].

CTAB has a lipid-like chemical structure; therefore, if CTAB adsorbs onto the algae cell surface it may enter the lipid extraction process where it may inadvertently contribute towards the total pool of recovered lipid. To determine the likelihood of this chemical carry over, known quantities of pure CTAB, 0.05–0.15 g, were put through the same total lipid extraction process as the microalgae and calculated as follows:

The mass of CTAB harvested (M_{CH}) was calculated using equation (2); where, M_c is the total quantity of CTAB added prior to harvesting, PA is the average percentage absorbance of CTAB onto the algal cell (calculated using method described in Section 2.3), N_0 is the total number of algae cells present in the culture, and N_1 is the total number of algae cells harvested:

Equation (2)

$$M_{CH} = M_c PA \left(\frac{N_1}{N_0} \right) \tag{2}$$

The total dry mass of cells harvested excluding CTAB (M_{CX}); where M_H is the total dry mass of the harvest was calculated using equation (3):

Equation (3)

$$MCX = MH - MCH \tag{3}$$

The total dry mass of lipid in the algal cells harvested by foam flotation excluding CTAB (M_{LCX}) was calculated using equation (4); where PL_{cent} is the average total lipid recovered from cells of the same age, and grown in the same culture as those harvested by foam flotation but harvested by centrifugation, the average was gained from three replicates (see Section 2.4):

Equation (4)

$$MLCX = P_{Lcent} MCX \tag{4}$$

The total mass of CTAB recovered by the lipid extraction process (M_{LC}) was calculated using equation (5); Where PL_{CTAB} is the average percentage of CTAB able to pass through the lipid extraction process. The average was gained from eight replicates starting with a known range of pure CTAB from 0.05 to 0.15 g to account for the potential increasing concentration of CTAB in the harvested foam:

Equation (5)

$$MLC = PLCTABMCH \qquad (5)$$

The predicted total lipid percentage ($PL_{(pred)}$) recovered from cells harvested by foam flotation was therefore calculated using equation (6):

Equation (6)

$$PL_{(pred)} = \frac{M_{LCX} + M_{LC}}{M_H} \qquad (6)$$

Fatty Acid Composition Analysis

Fatty acid composition was determined using a Carlo Erba Model Mega 5160 gas chromatograph (Carlo Erba, Milan, Italy). Ten milligrams of the freeze dried cells harvested by foam flotation and centrifugation on day 12 were placed into capped test tubes with heptadecanoic acid (C17:0) as an internal standard. Fatty acid methyl esters (FAMEs) were extracted by the one-step method of Graces and Mancha [39] (methanol: toluene: 2, 2 dimethoxypropane (DMP): sulphuric acid; 39:20:5:2, by volume). An injection volume of 1 µL was loaded onto a Supercool column (Sigma Aldrich) at 240 °C (30 m length × 0.25 mm ID, 0.25 µm film) with helium as the carrier gas. The temperature was programmed to ramp from 50 °C to 240 °C at 7 K min^{-1}. Fatty acids were identified by comparing the obtained retention times with that of known standards (37 component FAME mix, Supercool™). Fatty acid composition analysis was conducted in triplicate for each harvest method.

Statistical Analysis

Algal growth, concentration factors, total lipid, lipid classes and FAME contents for cells harvested by foam flotation and centrifugation were compared using analysis of variance (ANOVA). All percentage data were arcsine transformed prior to analysis. An alpha level of 0.05 was used to determine data significance.

RESULTS AND DISCUSSION

The cultures grew linearly until day 15, attaining a peak cell density of $4.44 \times 10^8 \pm 6.01 \times 10^7$ cm^3 followed by a significant decrease in cell density at day 21 to a cell density of $2.98 \times 10^8 \pm 3.69 \times 10^7$ cm^3 (p = 0.001).Fig. 2 illustrates the effect of culture age on the biomass concentration factor. The highest concentration factors – up to 306.89 ± 31.6 – which occurred on day 12 were significantly higher (p = <0.001) than the concentration factors gained for all other harvest days. The cultures appeared to enter a brief stationary phase by day 15 with cells beginning to aggregate and drop out of suspension. Danquah et al. [30] reported that microalgae harvested during a period of low growth had increased settling efficiencies compared with algae harvested during a high growth period. This was due to a reduction in the electrochemical stability of the cells within suspension measured as a decline in zeta potential during the low growth rate phase. This may explain why days 15, 18 and 21 yielded lower concentration factors of 111.29 ± 1.7, 250 ± 45.6, and 81.96 ± 6, respectively.

Changes in zeta potential have been linked to changes in the dissolved organic matter (DOM) pool within a culture. DOM concentration generally increases with culture age and can play an important role in promoting or inhibiting flocculation processes [31] and [40]. Zhang et al. [31] reported that, while harvesting using dissolved air flotation, DOM competed for the flocculent Al^{3+}, resulting in the stationary and declining growth phases requiring more Al^{3+} than the exponential growth phase. Unfortunately, the zeta potential could not be measured in the current study; however, Zhang et al. [31] demonstrated that the zeta potential of Chlorella zofingiensis varied significantly with growth phase, with the zeta potential declining from 20.6 ± 0.9, to 13.2 ± 3.0, and 12.2 ± 0.5 mV during the exponential, stationary, and declining growth phases respectively[31]. Using this information and that presented in Fig. 2, it can be concluded that the growth phase yielding the most advantageous flotation harvest efficiency in terms of concentration factor is during periods of high growth corresponding to low levels of DOM and a higher zeta potential (Fig. 2).

From the perspective of a commercial grower, these observations present an inconvenient, yet interesting paradox. If the grower is targeting lipid-derived bio products such as biodiesel, the period for optimal harvesting efficiency using foam flotation does not correspond to the period of maximal lipid yield (typically early stationary phase following a period of nitrogen deprivation [1]). However, if the grower is interested in biochemical synthesized during rapid growth, e.g. phycocyanin [41], then foam flotation would be appropriate. The inverse situation – characterized by a reduction in zeta potential, elevated DOM concentrations and a reduction in major cell surface functional groups (i.e. carboxyl, phosphate, and amine or hydroxyl groups; [31]) – will improve settling efficiency, and may therefore benefit flocculation and subsequent sedimentation. Nevertheless, a higher zeta potential may also increase the electrostatic interactions between CTAB and the negatively charged cell. CTAB can adsorb onto the surface of negatively charged particles [42], therefore increasing the hydrophobicity of the once hydrophilic solid–liquid interface and allowing efficient separation from the aqueous phase [11], [13] and [15]. CTAB may also create electrostatic interactions between the bubbles and the suspended particles thereby improving flotation [13]. When considering a multi-product operation, i.e. a bio refinery concept, this presents an intriguing challenge of how to select and optimize harvesting technologies and harvesting times to maximize product yield and therefore economic return. Such a situation will require detailed consideration of the techno economics of the harvesting process from a bio refining perspective. To our knowledge, this analysis has never been undertaken and further research is required.

To gain insight into which electrochemical interaction and hydrodynamic forces that dominate the foam flotation process, the adsorption of CTAB onto the algae cells was analyzed. The adsorption efficiency is influenced by a range of factors including cell age (and therefore zeta potential), cell density, and the surfactant concentration. It was assumed that the surfactant present in the supernatant was representative of that adsorbed onto the air/liquid interface, and that surfactant not present in the supernatant had adsorbed onto the algae cell. From Table 1 it can be seen that only $32.35 \pm 5.35\%$ of the total surfactant added was recovered from the supernatant after 1 h of exposure (Fig. 1, hypothesis 1). It can therefore be assumed that 32.35% of the CTAB added to the culture assumed a hydrophobic behavior, sticking the non-polar end of the molecule into the gas bubble [43]. It may therefore be deduced that the bulk of the surfactant ($67.65 \pm 5.35\%$) had been adsorbed onto the cells (Fig. 1, hypothesis

2). Surfactant adsorption strongly influences the efficiency of the foam flotation separation process as previously noted by Liu et al. [13] and Chen et al. [11]; however, no evidence of adsorption onto the cell surface was provided by these authors. The large variability in the dry mass yield (350 ± 120 mg L^{-1}) was due to flocs that had formed within the culture as it was over 21 days old. However, between each replicate the average dry weight of 343 ± 43 mg L^{-1} was used, and no significant difference (p = 0.402) was found in the noted concentration of surfactant removed in each replicate run. It can therefore be assumed that although flocks were present, the cell biomass in each replicate was evenly distributed. This data gives a good indication of the main electrochemical interaction and hydrodynamic forces that dominate the foam flotation process; however, further work is required to understand how CTAB adsorption changes with cell age and increases in cell density. Interestingly, it was also noted that cells harvested by CTAB-aided foam flotation had improved settling efficiencies over 30 min (14.3%) when compared to control settlement efficiencies (0.95%) (Data not shown). This is likely due to algal flocs being generated as a result of charge neutralization. Therefore, the concentration factor could be further increased post-harvest if a short settlement stage is added; however, further research is required to determine the effect of sedimentation post-harvest.

Table 1: The potential percentage adsorption of CTAB onto the algae cells as analyzed using a methyl orange colorimetric test. Data displayed is the mean of 4 replicates ± standard deviation

Supernatant removal stage	Surfactant mass fraction removed in supernatant (%)
V_1	29.05 ± 5.8
V_2	2.43 ± 0.74
V_3	0.87 ± 0.47
Total surfactant removed in supernatant	32.35 ± 5.35
Predicted surfactant adsorbed onto algae cell	67.65 ± 5.35

Lipid yields (mass fraction of dry cell weight) were within a similar range to those reported in other studies [44] and [45]; however, it should be noted that the cultures in the current study were not deliberately nitrogen starved and therefore lipid biosynthesis was not optimized. The total lipid extracted from the dry biomass harvested by foam flotation varied between a minimum of 14.5 ± 0.47% on day 3, to a maximum of 17.63 ± 0.53% on day 6. For the cells harvested by centrifugation the minimum recovery was 8.3 ± 0.16% on day 3 with a maximum of 11.4 ± 0.47% on day 21 (Fig. 3). Mean lipid yield from all foam flotation harvests was 15.4 ± 0.77% which was significantly higher (p = <0.001) than yields from centrifugation (9.9 ± 0.56%). No significant relationship was observed between total lipid content and growth stage for algae harvested by either method. Although the nutrient levels were not monitored throughout the growth cycle, the lack of any significant increase in lipid content of cells harvested by centrifugation (p = 0.092) suggests the cells were not nitrogen limited. This supports the theory that a reduction in the electrochemical stability of the cells within suspension caused the cells to aggregate and drop out of suspension [30], thereby resulting in the significant decline in cell density at day 21 (Fig. 2), and reducing the harvesting efficiency towards the end of the growth phase at days 15, 18 and 21. Unexpectedly, the results indicated that CTAB-aided foam flotation significantly increased the extractable lipid fraction.

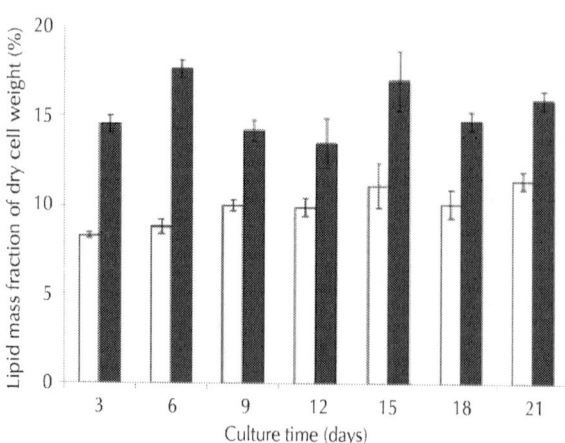

Figure. 3: The lipid mass fraction of dry cell weight (% mean ± standard deviation) extracted from cells harvested by centrifugation (grey) and foam flotation (black).

Despite the harvested cells being washed and freeze dried prior to lipid extraction, a significant amount of surfactant may still have been adsorbed to the cell surface. Therefore, the capacity for CTAB to be carried over into the total lipid extraction procedure was considered. Up to 79.4 ± 4.0% of CTAB by weight was recovered from the total lipid extraction, with an average of 76.6 ± 8.74% recorded. The effect of CTAB on the extractable lipid pool was calculated based on total lipid recovered from flotation and centrifuge harvested cells, CTAB adsorption efficiency, and the proportion of CTAB recoverably from the total lipid extraction process.

Table 2 shows that the predicted values for the total lipid yields from cells harvested by foam flotation (calculated using equations (2), (3), (4), (5) and (6)) were close to those actually measured ($p = 0.896$), suggesting that the higher yields from foam flotation harvested cells were due, in part, to CTAB adsorption and subsequent carry over into the extraction process. Although a number of papers have focused on foam flotation [11], [13], [14] and [15], to our knowledge the effect of surfactant adsorption on lipid content has never been noted.

Table 2: The predicted and actual total recovered lipid from cells harvested by foam flotation. The predicted values were calculated by adding the lipids content of cells harvested by centrifugation with the calculated mass of CTAB that was removed from the total lipid extraction process, using equations (2), (3), (4), (5) and (6)

Culture age	Predicted	Gained
3	16.6	14.5
6	11.3	17.6
9	15.7	14.2
12	18.9	13.4
15	18.0	17.0
18	15.6	14.3
21	16.9	15.9

It has however been found that the electrostatic interactions that aid flotation may also compromise the integrity of the cell wall, which may potentially result in cell lysis [46]. It has been suggested that

CTAB effects gross membrane damage through protein denaturation, causing the cell membrane to rupture [46] and [47]. It is feasible that the increased lipid yields from foam flotation harvested cells may also be due in part to an increase in the concentration of phospholipids liberated due to membrane deterioration. To explore this, the lipid fractions of cells harvested by foam flotation and centrifugation were investigated.

The total lipid pool is composed primarily of neutral lipid in the form of energy storage bodies, as well as glycol and phospholipids within structural membranes. Within the literature the total lipid content of microalgae is commonly quoted as an indication of its appropriateness as a biodiesel feedstock [1]. However, not all lipid fractions can be easily esterified into biodiesel; therefore a high quoted total lipid may be misleading. Solid phase extraction was used to determine the crude composition of the total lipid extract from which information on the effect of harvesting technique on the biodiesel production potential and the impact of CTAB on the total lipid recovered could be determined. The extracted lipids from cells harvested by foam flotation and centrifugation consisted mainly of neutral lipids (Table 3), contributing $60.1 \pm 12.5\%$ and $58.3 \pm 12.2\%$ of the total lipid pool respectively. There was no significant difference between the neutral lipid fraction ($p = 0.739$), and the free fatty acid fraction (FFA) ($p = 0.085$) when comparing the lipids recovered from either harvest method. However, a significantly higher phospholipid content of $23.6 \pm 5.4\%$ was extracted from microalgae harvested by foam flotation ($p = 0.008$) compared with centrifugation ($16.1 \pm 12.8\%$). As all harvested cultures were grown under the same environmental conditions it is unlikely that foam flotation harvested cells would have contained a greater titer of phospholipids.

Table 3: Percentage of lipid class mass fraction with respect to total extracted lipid. Data displayed is the mean of 4 replicates ± standard deviation

Lipid fraction	Foam flotation	Centrifugation	CTAB mixture
Neutral lipid	60.1 ± 12.5	58.3 ± 12.2	37.8 ± 14.5
Free fatty acids	16.1 ± 8.9	25.5 ± 4.6	1.7 ± 1.6
Phospholipids	23.6 ± 5.4	16.1 ± 12.8	0.4 ± 0.8

CTAB that had been put through the modified Folch method was also separated by solid phase extraction. The CTAB in chloroform was applied to the separation column under vacuum, which should have left the full mass of CTAB on the column; however, $27.6 \pm 17.1\%$ was removed during the application process. From Table 3 it can be seen that the majority of the CTAB was eluted as a neutral lipid. There was no significant difference in the percentage of neutral lipids when comparing cells harvested by foam flotation and centrifugation; therefore, it is unlikely that CTAB would have affected this cell fraction. Only minor fractions of CTAB were recovered as FFA, and phospholipid, with $25.61 \pm 6.05\%$ of CTAB remaining on the column. This data suggests that the adsorbed CTAB did not significantly affect the composition of the lipid fraction. Therefore, CTAB must have increased the amount of phospholipids available for extraction due to the solubilisation of the phospholipid cell membrane [48]; this may also have contributed to the increased total recovered lipid when harvesting by foam flotation.

The neutral lipids found within microalgae extracts are mainly comprised of triacylglycerol's (TAGs) [49]. Although both polar and neutral lipids can be converted to biodiesel, neutral lipids are the desired fraction as TAGs are easily Trans esterified to biodiesel [50]. FFA can also be converted after esterification [25]; however, Van Gerpen [51] reported that phosphorus containing compounds in the crude lipid oil did not convert into the methyl esters, which may cause problems during conversion and combustion processes [51] and [52].

However, whilst not necessarily valuable for biodiesel production, there are established and growing markets for certain phospholipids and their by-products [53] that may form important outputs as part of an algae bio refinery. CTAB is used as a food grade chemical for the extraction of pigments from red beet; therefore procedures to ensure that the product is fit for human/animal consumption are established [48]. Phospholipids can also be recycled as sources of nitrogen and phosphorus for microalgae cultivation, which could significantly reduce the operational production costs [54].

The effect of foam flotation on the fatty acid profile was investigated (Table 4). It was important to characterize the fatty acid profile as this can dramatically affect the quality of the biodiesel product [26] and inform the economics of a bio refinery producing lipid-based high value products. No significant difference was found between the quantity of

FAMEs gained from cells harvest by centrifugation or foam flotation (p = 0.609), confirming that the adsorbed CTAB did not significantly affect the neutral lipid fraction. The total tranesterifiable lipid for cells harvested by centrifugation was 6.4 ± 1.3% dry weight (DW) and 5.6 ± 0.3% for cells harvested by foam flotation. However, discernible changes within the FAME profiles were noted. Significantly higher yields of the monounsaturated fatty acid (MUFA) oleic acid (C18:1n9c) were recovered from cells harvested by foam flotation (5.1 ± 0.133% DW) compared to cells harvested by centrifugation (2.17 ± 0.08% DW) (p = <0.001). Significantly greater yields of total MUFA (9.7 ± 0.15% DW) (p = <0.001) and saturated fatty acids (SFA) (6.4 ± 0.05% DW) (p = 0.006) were also recovered from foam flotation harvested cells (Table 5). In terms of biodiesel quality, higher proportions of MUFA and SFA are desirable as they increase the fuels› energy yield, cetin number, and also improve the oxidative stability [44]. Surprisingly, cells harvested by centrifugation had higher yields of total polyunsaturated fatty acids (PUFA) at 32.5 ± 0.48% DW compared to 25.3 ± 0.54% DW for cells harvested by foam flotation (p = 0.001); including higher yields of linoleic (C18:2n6c) (p = 0.001) and linolenic acids (C18:3n3) (p = 0.001) (Table 4). There was no significant difference between the C18 series between either harvest method (p = 0.084). Knot he [55] stated that politic, stearic, oleic, and linolenic acids are the most common fatty acids present in biodiesel. These components equate to 24.7 ± 0.46% for centrifugation and 23.3 ± 0.30% for foam flotation; there was no significant difference between the harvesting methods (p = 0.091) (Table 5).

Table 4: Selected fatty acid methyl ester profiles with respect to the dry mass fraction of total fatty acids (%). Data are the mean of 3 replicates. A broader suite of FAMES, including C18, were detected but not presented here. Total C18 series is the sum of all C18 FAMEs identified including those not listed in the table

Fatty acid methyl ester	Centrifugation	Foam flotation
Caprylic (C8:0)	0.04	0.05
Capric (C10:0)	0.03	0.05
Lauric (C12:0)	0.24	0.21
Myristic (C14:0)	1.27	1.77

Pentadecanoic (C15:0)	0.12	0.18
Palmitic (C16:0)	1.76	1.52
Stearic (C18:0)	0.30	0.69
Elaidic (C18:1n9t)	4.54	4.53
Oleic (C18:1n9c)	2.17	5.16
Linoleic acid (C18:2n6c)	22.30	17.48
Linolenic acid (C18:3n3)	10.20	8.14
Arachidic (C20:0)	2.23	1.68
cis-11-Eicosenoic (C20:1)	0.12	0.10
Behenic (C22:0)	0.10	0.24
Total C18 series	41.12	38.85

Table 5: Composition and mass fraction of FAME (%) harvested by centrifugation and foam flotation. Data expressed as means of 3 replicates ± standard deviation. Saturated Fatty acids = C8:0, C10:0, C12:0, C14:0, C15:0, C16:0, C18:0, C20:0, C22:0; Monounsaturated fatty acids = C18:1n9t, C18:1n9c; Polyunsaturated fatty acids = C18:2N6c, C18:3n3; Dominant biodiesel components = C18:0, C18:1n9c, C18:2n6c [25]

Properties	Centrifugation	Foam flotation
Saturated fatty acids (% of total FAME)	6.0 ± 0.04	6.4 ± 0.05
Monounsaturated fatty acids (% of total FAME)	6.7 ± 0.09	9.7 ± 0.15
Polyunsaturated fatty acids (% of total FAME)	32.5 ± 0.48	25.3 ± 0.54
Dominant biodiesel components (% of total FAME)	24.7 ± 0.92	23.3 ± 0.30
Total fame content (%CDW)	6.4 ± 1.27	5.6 ± 0.26

Lee et al. [32] tested the effect of three different flocculating methods: pH adjustment, treatment with aluminum sulphate, and treatment with Pestan (a microbial flocculants), on the lipid content of Botryococcus braunii. It was found that the total lipid content was unaffected by the harvest method; however, no investigation into the fatty acid profile was carried out. Borges et al. [28] also found no significant difference for the total microalgae lipid content with respect to harvest method when comparing anionic and cationic polyacrylamide flocculants; however, the fatty acid profile differed significantly between different flocculants. It would therefore appear that the choice of harvest method can greatly affect lipid product quantity and quality.

CONCLUSIONS

Harvesting of Chlorella sp. by foam flotation is most effective during phases of active culture growth, suggesting that foam flotation may prove particularly advantageous for species that synthesize desirable biochemical during active growth, but not as beneficial necessarily for species cultured specifically for biodiesel production. A greater quantity of lipid was recovered when biomass was harvested by foam flotation as opposed to centrifugation. This study is the first to investigate the effect of CTAB-aided foam flotation harvesting on lipid content and fatty acid profiles. The improved lipid recovery occurred due to a combination of an increase in the total extractable lipid caused by the solubilisation of the phospholipid bilayer by the surfactant CTAB, and also a proportion of the CTAB dose becoming adsorbed onto the cell and entering the lipid extraction process. Foam flotation resulted in a predominance of saturated and monounsaturated fatty acids within the fatty acid profile, which provide many favorable features for biodiesel production. Foam flotation is an advantageous microalgae harvesting technique and a full techno economic analysis in relation to microalgae bio refining is greatly needed.

ACKNOWLEDGEMENTS

This work was supported by an Engineering and Physical Sciences Research Council (EPSRC), EP/P502624/1 doctoral studentship to Thea Coward. The authors are grateful to S. Latimer, P. Carrick, H. Redden, S. Conlan, R. Taylor, J. Rand, and D. Whitaker for assistance during the course of the work.

REFERENCES

1.   Chisti Y. Biodiesel from microalgae. Biotechnology Adv2007; 25(3):294e306.

2. Christenson L, Sims R. Production and harvesting of microalgae for wastewater treatment, biofuels, and bio products. Biotechnology Adv 2011; 29(6):686e702.

3. Demirbas A, Fatih Demirbas M. Importance of algae oil as a source of biodiesel. Energy Convers Manage 2011; 52(1):163e70.

4. Perez-Garcia O, Escalante FME, de-Bashan LE, Bashan Y. Heterotrophic cultures of microalgae: metabolism and potential products. Water Res 2011; 45(1):11e36.

5. Uduman N, Qi Y, Danquah MK, Forde GM, Hoadley A. Dewatering of microalgal cultures: a major bottleneck to algae-based fuels. J Renew Sustain Energy 2010; 2(1):1e15.

6. Schenk P, Thomas-Hall S, Stephens E, Marx U, Mussgnug J,Posten C, et al. Second generation biofuels: high-efficiency microalgae for biodiesel production. Bioenergy Res 2008; 1(1):20e43.

7. Olguín EJ. Dual purpose microalgaeebacteria-based systems that treat wastewater and produce biodiesel and chemical products within a biorefinery. Biotechnol Adv 2012; 30(5):1031e46.

8. Slade R, Bauen A. Micro-algae cultivation for biofuels: cost, energy balance, environmental impacts and future prospects. Biomass Bioenergy 2013; 53:29e38.

9. Dassey AJ, Theegala CS. Harvesting economics and strategies using centrifugation for cost effective separation of microalgae cells for biodiesel applications. Bioresour Technol 2013; 128:241e5.

10. Coward T, Lee JGM, Caldwell GS. Development of a foam flotation system for harvesting microalgae biomass. Algal Res 2013; 2(2):135e44.

11. Chen YM, Liu JC, Ju YH. Flotation removal of algae from water. Colloid Surf B 1998; 12(1):49e55.

12. Csordas A, Wang JK. An integrated photobioreactor and foam fractionation unit for the growth and harvest of Chaetoceros spp. in open systems. Aquacult Eng 2004; 30(1e2):15e30.

13. Liu JC, Chen YM, Ju YH. Separation of algal cells from water by column flotation. Separ Sci Technol 1999; 34(11):2259e72.

14. Phoochinda W, White DA. Removal of algae using froth flotation. Environ Technol 2003; 24(1):87e96.

15. Phoochinda W, White DA, Briscoe BJ. An algal removal using a combination of flocculation and flotation processes. Environ Technol 2004; 25(12):1385e95.

16. Wiley PE, Brenneman KJ, Jacobson AE. Improved algal harvesting using suspended air flotation. Water Environ Res 2009; 81(7):702e8.

17. Brennan L, Owende P. Biofuels from microalgae e a review of technologies for production, processing, and extractions of biofuels and co-products. Renew Sustain Energy Rev 2010; 14(2):557e77.

18. Henderson RK, Parsons SA, Jefferson B. The impact of differing cell and algogenic organic matter (AOM) characteristics on the coagulation and flotation of algae. Water Res 2010; 44(12):3617e24.

19. Wiley PE, Campbell JE, McKuin B. Production of biodiesel and biogas from algae: a review of process train options. Water Environ Res 2011; 83(4):326e38.

20. Edzwald JK. Dissolved air flotation and me. Water Res 2010; 44(7):2077e106.

21. Henderson RK, Parsons SA, Jefferson B. Surfactants as bubble surface modifiers in the flotation of algae: dissolved air flotation that utilizes a chemically modified bubble surface. Environ Sci Technol 2008; 42(13):4883e8.

22. Chung Y, Choi YC, Choi YH, Kang HS. A demonstration scaling-up of the dissolved air flotation. Water Res 2000; 34(3):817e24.

23. DeSousa SR, Laluce C, Jafelicci M. Effects of organic and inorganic additives on flotation recovery of washed cells of Saccharomyces cerevisiae resuspended in water. Colloid Surf B 2006; 48(1):77e83.

24. Cho S-H, Kim J-Y, Chun J-H, Kim J-D. Ultrasonic formation of nanobubbles and their zeta-potentials in aqueous electrolyte and surfactant solutions. Colloid Surf A 2005; 269(1e3):28e34.

25. Doan TTY, Sivaloganathan B, Obbard JP. Screening of marine microalgae for biodiesel feedstock. Biomass Bioenergy 2011; 35(7):2534e44.

26. Gouveia L, Oliveira AC. Microalgae as a raw material for biofuels production. J Ind Microbiol Biotechnol 2009; 36(2):269e74.

27. Knothe G. Dependence of biodiesel fuel properties on the structure of fatty acid alkyl esters. Fuel Process Technol 2005; 86(10):1059e70.

28. Borges L, Moron-Villarreyes JA, D'Oca MGM, Abreu PC. Effects of flocculants on lipid extraction and fatty acid composition of the microalgae Nannochloropsis oculata and Thalassiosira weissflogii. Biomass Bioenergy 2011; 35(10):4449e54.

29. Becker EW. Microalgae: biotechnology and microbiology. Cambridge, UK: Cambridge University Press; 1994. p. 293.

30. Danquah MK, Gladman B, Moheimani N, Forde GM. Microalgal growth characteristics and subsequent influence on dewatering efficiency. Chem Eng J 2009; 151(1e3):73e8.

31. Zhang X, Amendola P, Hewson JC, Sommerfeld M, Hu Q. Influence of growth phase on harvesting of Chlorella zofingiensis by dissolved air flotation. Bioresour Technol 2012; 116: 477e84.

32. Lee SJ, Kim SB, Kim JE, Kwon GS, Yoon BD, Oh HM. Effects ofharvesting method and growth stage on the flocculation of the green alga Botryococcus braunii. Lett Appl Microbiol 1998; 27(1):14e8.

33. Borowitzka MA. Commercial production of microalgae: ponds, tanks, tubes and fermenters. J Biotechnol1999; 70(1e3):313e21.

34. Al-Mashhadani MKH, Bandulasena HCH, Zimmerman WB. CO2 mass transfer induced through an airlift loop by a microbubble cloud generated by fluidic oscillation. Ind Eng Chem Res 2012; 51(4):1864e77.

35. Wang L, Langley D. Determining cationic surfactant concentration. Ind Eng Chem Prod RD 1975; 14(3):210e2.

36. Folch J, Lees M, Sloane-Stanley GH. A simple method for the isolation and purification of total lipids from animal tissues. J Biol Chem 1957; 226(1):497e509.

37. Kaluzny MA, Duncan LA, Merritt MV, Epps DE. Rapid separation of lipid classes in high yield and purity using bonded phase columns. J Lipid Res 1985; 26(1):135e40.

38. Peplow A, Balaban M, Leak F. Lipid composition of fat trimmings from farm-raised alligator. Aquaculture 1990; 91(3e4):339e48.

39. Garces R, Mancha M. One-step lipid extraction and fatty acid methyl esters preparation from fresh plant tissues. Anal Biochem 1993; 211(1):139e43.

40. Henderson R, Parsons SA, Jefferson B. The impact of algal properties and pre-oxidation on solid-liquid separation of algae. Water Res 2008; 42(8e9):1827e45.

41. Soni B, Kalavadia B, Trivedi U, Madamwar D. Extraction, purification and characterization of phycocyanin from Oscillatoria quadripunctulata isolated from the rocky shores of Bet-Dwarka, Gujarat, India. Process Biochem 2006; 41(9):2017e23.

42. Paria S, Khilar KC. A review on experimental studies of surfactant adsorption at the hydrophilic solid water interface. Adv Colloid Interface Sci 2004; 110(3):75e95.

43. Timmons MB, Ebeling JM, Wheaton FW, Summerfelt ST, Vinci BJ. Recirculating aquaculture. 2nd Ed. New York: Cayuga Aqua Ventures; 2002.

44. Renaud S, Parry D, Thinh L-V. Microalgae for use in tropical aquaculture I: gross chemical and fatty acid composition of twelve species of microalgae from the Northern Territory, Australia. J Appl Phycol 1994; 6(3):337e45.

45. Yoo C, Jun S-Y, Lee J-Y, Ahn C-Y, Oh H-M. Selection of microalgae for lipid production under high levels carbon dioxide. Bioresour Technol 2010; 101(1):71e4.

46. Simoes M, Pereira M, Machado I, Sim ~ oes L, Vieira M. ~ Comparative antibacterial potential of selected aldehydebased biocides and surfactants against planktonic Pseudomonas fluorescens. J Ind Microbiol Biotechnol 2006; 33(9):741e9.

47. Gilbert P, Allison DG, McBain AJ. Biofilms in vitro and in vivo:do singular mechanisms imply cross-resistance? J Appl Microbiol 2002; 92: 98e110.

48. Thimmaraju R, Bhagyalakshmi N, Narayan MS,Ravishankar GA. Food-grade chemical and biological agents permeabilize red beet hairy roots, assisting the release of betalaines. Biotechnol Prog 2003; 19(4):1274e82.

49. Hu Q, Sommerfeld M, Jarvis E, Ghirardi M, Posewitz M, eibert M, et al. Microalgal triacylglycerols as feedstocks for

biofuel production: perspectives and advances. Plant J 2008; 54(4):621e39.

50. Ma F, Hanna MA. Biodiesel production: a review. Bioresour Technol 1999; 70(1):1e15.

51. Van Gerpen J. Biodiesel processing and production. Fuel Process Technol 2005; 86(10):1097e107. [52] Pruvost J, Van Vooren G, Cogne G, Legrand J. Investigation

52. Of biomass and lipids production with Neochloris oleoabundans in photobioreactor. Bioresour Technol 2009; 100(23):5988e95.

53. Alonso DL, Belarbi E-H, Rodriguez-Ruiz J, Segura CI, Gimenez A. Acyl lipids of three microalgae. Phytochemistry 1998; 47(8):1473e81.

54. Rosch C, Skarka J, Wegerer N. Materials flow modeling of nutrient recycling in biodiesel production from microalgae. Bioresour Technol 2012; 107(0):191e9.

55. Knothe G. "Designer" biodiesel: optimizing fatty ester composition to improve fuel properties. Energy Fuel 2008; 22(2):1358e64.

Chapter 5

Marginal Cost of Delivering Switchgrass Feedstock and Producing Cellulosic Ethanol at Multiple Biorefineries

Mohua Haque[a], Francis M. Epplin[b], Jon T. Biermacher[a], Rodney B. Holcomb[c], and Philip L. Kenkel[d,]

[a]Agricultural Division, The Samuel Roberts Noble Foundation, 2510 Sam Noble Parkway, Ardmore, OK 73401, United States

[b]Department of Agricultural Economics, Oklahoma State University, 316 Ag Hall, Stillwater, OK 74078-6026, United States

[c]Department of Agricultural Economics, Oklahoma State University, 114 Food & Agricultural Products Center, Stillwater, OK 74078-6026, United States

[d]Department of Agricultural Economics, Oklahoma State University, 516 Ag Hall, Stillwater, OK 74078-6026, United States

ABSTRACT

Limited information is available regarding the change in cost to deliver dedicated energy crop feedstock as the quantity of required feedstock increases. The objective is to determine the marginal cost to produce and deliver switchgrass feedstock to biorefineries. A mathematical programming model that includes 77 production regions (Oklahoma counties), monthly feedstock requirements, integer activities for harvest machines and integer activities for each of 16 potential biorefinery locations was constructed. The model was initially solved for a single biorefinery. The number of plants was incremented by one and the model resolved until nearly 10% of the cropland and improved pasture land was converted to switchgrass. The estimated cost to deliver 1.0 Mg of feedstock to a single 189 dam^3 y^{-1} capacity biorefinery is 55 $. The cost to deliver feedstock increases as additional biorefineries are constructed and the cost for the ninth biorefinery of 87 $ Mg^{-1} is 58% greater than the cost to deliver to the first biorefinery. The cost difference is primarily due to differences in transportation cost. Initial cellulosic biorefineries will have an opportunity for establishing a feedstock cost advantage by carefully selecting land for conversion to switchgrass and by negotiating long term leases.

INTRODUCTION

In 2007, the U.S. Energy Independence and Security Act (EISA) was passed by the U.S. Congress and signed by President Bush to encourage production of biofuel from cellulosic feedstock. EISA mandated that if produced, 136 hm^3 of renewable fuel should be used in the nation's fuel supply by the year 2022, including 61 hm^3 from advanced cellulosic feedstocks. To meet this biofuel mandate it is expected that the majority will be produced from lignocellulosic feedstocks such as agricultural crop residues, waste products, woody biomass and dedicated energy crops [1]. The Department of Energy has proposed that dedicated energy crops such as switchgrass produced on marginal lands could provide a substantial quantity of low cost feedstock [2]. Switchgrass may be a viable alternative but currently infrastructure for production, harvest, storage and transportation of switchgrass biomass does not exist [3]. Given an expected conversion rate of 0.35 m^3 Mg^{-1}, a 189 dam^3 y^{-1} biorefinery would require 1.483 Gg d^{-1} of switchgrass biomass [4]. For an average harvestable yield of 10 Mg ha^{-1} such a biorefinery would require production from 148 ha d^{-1}[5].

As the number of biorefineries in a region increases the competition for land to produce feedstock will increase. As a result, some feedstock will have to be produced on less productive land and transported greater distances, both of which will increase the average cost. Timmons [6] reported that if less productive land is used to grow switchgrass biomass, biofuel production costs will increase. In addition, research has reported that transportation costs would comprise a large percentage of biomass feedstock delivery cost [7],[8], [9] and [10]. Several studies have used a Geographic Information System (GIS) to identify potential facility locations for biofuel production as biomass feedstock is geographically dispersed and the location of a biofuel plant influences transportation costs [10] and [11]. Graham et al. [12] estimated the marginal cost of delivering biomass (wood chips) from different regions of the state of Tennessee and found that the marginal costs of delivered chips varied by both facility location and facility demand. However, limited information exists regarding the expected increase in cost to deliver feedstock to the biorefinery as the number of biorefineries increases.

A number of studies have evaluated the farm gate costs of producing switchgrass [13], [14], [15] and [16]. A few studies have evaluated the cost to deliver switchgrass feedstock to a single biorefinery or facility, and the results are mixed [3], [7], [8], [17], [18] and [19]. For instance, Brechbill et al. [17] estimated that the total cost to deliver switchgrass biomass a distance of 60 km in Indiana ranged between 80 and 90 $ Mg^{-1}. Vadas, Barnett, and Undersander [18] estimated a cost of 77 $ Mg^{-1} to deliver switchgrass biomass to a biorefinery in Wisconsin with land and transportation cost comprising 44% of the total cost. Studies conducted in Illinois and Iowa found production, harvest, storage and transportation costs for switchgrass biomass delivered to a single biorefinery were 98 and 125 $ Mg^{-1}, respectively [7] and [8]. Both of these studies also found that land, production and transportation costs of switchgrass were 57% of the total cost considered. In contrast, Epplin et al. [3] estimated switchgrass total delivery cost of 54 $ Mg^{-1} for a biorefinery located in Oklahoma and reported that land, production and transportation cost comprised 65% of the total cost. Graham, English, and Noon [13] estimated switchgrass delivered feedstock costs ranging from 33 to 55 $ Mg^{-1} across eleven US states to supply a facility requiring 100 Gg y^{-1}. They found that feedstock transportation costs were greater for larger facilities. These studies focused on modeling cellulosic biomass logistics issues (i.e., establishment, fertilizer management, harvest, transportation, storage) for a single biorefinery. Fewer studies have evaluated cost consequences of multiple biorefineries competing for land to produce feedstock [20].

The U.S. Environmental Protection Agency (EPA) projected that it will be economically feasible to produce 3 hm^3 of cellulosic ethanol from switchgrass by 2022 [2]. In addition, EPA projected that the majority of the switchgrass would be grown in Oklahoma, perhaps in part because of the variety of land resources and the relatively low opportunity cost of land in the state. The state has 60,300 km^2 in native prairie grass, 19,000 km^2 in improved pasture, 4000 km^2 in the federal government's Conservation Reserve Program, and 34,000 km^2 of harvested cropland.

The objective is to determine the marginal cost to produce and deliver switchgrass feedstock to biorefineries. The case study includes Oklahoma's 77 counties as production regions. Cost estimates are produced under the assumptions that (1) no more than 10% of a county's cropland and improved pasture land may be bid from current use and converted to switchgrass; (2) each biorefinery has a capacity of 189 dam^3 y^{-1}; and (3) switchgrass biomass is converted to ethanol at a rate of 0.35 m^3 Mg^{-1}. The number of potential biorefineries is incrementally increased in the model from one until nearly 10% of the state's cropland and improved pasture land is converted. This enables an estimate of the economic consequences of expanding the industry into less favorable locations that would require transportation of biomass from greater distances and production of switchgrass on lands with lower expected yields resulting in greater costs per delivered Mg.

Information gathered from this research provides an estimate of the changes in marginal cost to deliver feedstock as the number of biorefineries increase to produce the mandated levels of biofuels from a potential dedicated cellulosic energy crop. Prior studies have argued that the cost to produce biofuel will be lower from the nth plant than from initial biorefineries as engineers fine-tune the feedstock-to-biobased products production system [4] and [20]. However, these studies have ignored the potential increase in feedstock cost for the nth relative to the initial biorefinery.

MODELING, DATA AND ASSUMPTIONS

This study used a multi-region, multi-period, mixed integer mathematical programming model similar to models used in previous studies [20], [22], [23], [24] and [25]. The model is designed and solved to determine the cost to procure, harvest, store and transport a flow of switchgrass biomass to an optimally located set of biorefineries (with the number of biorefineries ranging from one to nine), the area and quantity of switchgrass harvested by county and the number of harvest machines. The model also determines optimal number of harvest units, number of harvest machines and average investment in harvest machines. Binary variables are included to enable the model to determine the most economical plant locations. Integer variables are used to determine the optimal number of harvest machines. The model was solved using the generalized algebraic modeling system (GAMS) with the CPLEX solver. The model includes about 42,200 activities and 9000 equations.

Since a spot market for switchgrass feedstock does not exist, a vertically integrated system similar to that used by timber industries [24], [26], [27] and [28] is assumed for modeling purposes. Modeling was based on the assumption that switchgrass production, harvest, storage and transportation would be centrally managing. The model is initially solved to determine the cost to deliver feedstock and the breakeven ethanol price for a single biorefinery. The number of plants is incremented by one and the model resolved until nearly 10% of the cropland and improved pasture land in the 77 counties is converted to switchgrass. It is assumed that plant construction and establishment of switchgrass feedstock occurs in year zero. Activities from year 1 through 20 are assumed to be identical meaning that the annual net benefit is modeled as an annuity. Parameters and parameter values used to estimate the model are presented in Table 1.

Table 1: Parameters and parameter values

Item	Item value
Biorefinery	
Product	Ethanol
Conversion technology	Gasification–fermentation
Each plant capacity, $dam^3\ y^{-1}$	189
Capital investment for each plant, M$	379
Operation & maintenance cost, $\$\ m^{-3}$	260
Conversion rate, m^3 of ethanol Mg^{-1}	0.35
Plant life, years	20
Feedstock and land	
Cropland rental lease rate, $\$\ ha^{-1}\ y^{-1}$	148
Improve pasture land lease rate, $\$\ ha^{-1}\ y^{-1}$	99
Proportion of cropland available for leasing, %	10
Proportion of improved pasture land available for leasing, %	10
Price of nitrogen fertilizer, $\$\ kg^{-1}$	1.0
Price of phosphorus fertilizer, $\$\ kg^{-1}$	1.2
Amortized establishment cost on cropland, $\$\ ha^{-1}\ y^{-1}$	68.8
Amortized establishment cost on improve pasture land, $\$\ ha^{-1}\ y^{-1}$	61.8
Maintenance cost, $\$\ ha^{-1}\ y^{-1}$	10.2
Diesel fuel price, $\$\ m^{-3}$	680
Biomass field storage cost, $\$\ Mg^{-1}$	2.2
Other assumptions	
Potential biorefinery locations, number	16
Potential switchgrass production regions, number	77

Biorefinery and Feedstock Supply Locations

Sixteen counties were selected as potential biorefinery locations: Pontotoc, Washington, Canadian, Garfield, Okmulgee, Payne, Blaine, Carter, Grady, Kay, Woods, Comanche, Custer, Jackson, Texas and Woodward. These potential biorefinery locations were selected based on biomass feedstock relative density and availability of accessible-road infrastructure. A map of Oklahoma counties showing the 77 potential production regions and 16 potential biorefinery locations is presented in Fig. 1.

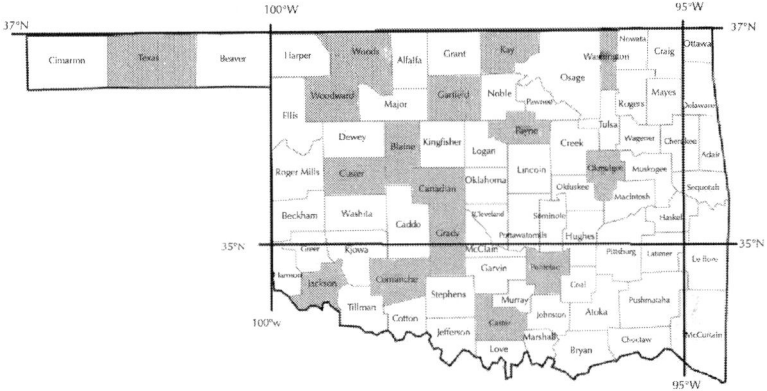

Figure1: Map of Oklahoma counties showing 77 potential switchgrass production regions and 16 potential biorefinery locations (in gray).

ACQUISITION OF LAND USE AND BIOMASS YIELD

The quantity of cropland and improved pasture land for each Oklahoma county was determined from data reported by the Census of Agriculture [29]. By assumption, land available for conversion from current use to switchgrass was restricted to be no more than 10% of each county's cropland and 10% of each county's pasture land. A biorefinery or group of biorefineries (system) could engage in long term leases with land owners and produce switchgrass on both cropland and improved pasture land. Average 2006–2010 non-irrigated cropland and improved pasture land cash rental rate for Oklahoma ranged from 67 to 69 $ ha^{-1} and 21–27 $ ha^{-1}, respectively [30]. Average long term lease rates that would be required to bid up to 10% of cropland and improved pasture land from current use are unknown. For comparison the average Oklahoma Conservation Reserve Program (CRP) rental payment as of May 2012 for 10-year land leases was 84 $ ha^{-1}[31]. To compensate for the uncertainty arising from the length of the potential lease and to allow for an increase in land rental rates in response to the potential 10% increase in quantity demanded, rates of 148 $ ha^{-1} and 99 $ ha^{-1} were used for cropland and improved pasture land, respectively [25].

Switchgrass yields for each of the 77 counties for both cropland and improved pasture land were based on estimates produced by Basnet et al. [32]. For this study, only average county yields are considered since data regarding yield variability are limited. If yields are normally distributed, total production from the leased land would be insufficient to meet the needs of the biorefinery in approximately half of the production years. If biomass was not available from other sources, such as from production in prior years or from land not leased, the biorefinery would have to be idled for a period of time depending on the production shortage. The expected yield and fertilizer requirements are adjusted depending on month of harvest [33] and [34]. Switchgrass harvested in September and October produces greater expected yield and has lower requirement for fertilizer because of nutrient recycling compared to switchgrass harvested in mid-season in July and August. It is also expected that if switchgrass is left to stand in the field there will be dry matter losses of 5% per month from November through March [17], [20], [32] and [35].

Harvest, Storage, and Transportation

The model was constructed based on the assumption that switchgrass could be harvested once per year and that the harvest window could extend from July through March enabling a just-in-time harvest and delivery system for most of the year [34]. Harvest operations include mowing, raking, baling and stacking. Days suitable for mowing and baling switchgrass depend on the weather [35]. Distributions of suitable mowing and baling days by month for each of the 77 counties were determined by Hwang et al. [36]. For instance, based on the work day distributions, in 19 of 20 years in Grady County at least 16 days will be suitable for mowing and at least five days will be suitable for baling in October. Harvest cost was estimated based on the integrated harvest unit concept developed by Thorsell et al. [37]. Harvest units were included as integer variables in the model and separated into mowing and baling (rake-bale-stack) units. Mowing units include a self-propelled windrower (142 kJ s^{-1}) equipped with a 16 foot rotary header and a laborer. The raking-baling-stacking harvest unit includes three wheel rakes, three 41 kJ s^{-1} tractors; three balers, three 149 kJ s^{-1} tractors; a field transporter; and seven laborers. For large volume, and current forage harvest technologies, to collect for field storage and transport substantial distances, large rectangular bales is the least-cost system for harvesting biomass from switchgrass in Oklahoma [38]. The balers are designed to form 1.22 m × 1.22 m × 2.44 m rectangular solid bales [37].

Harvested feedstock may be transported by truck just-in-time after baling or stored in the field until needed and transported later. Field storage stacks were assumed to be covered by a plastic tarp at a cost of 2.20 \$ Mg^{-1}[22]. Dry matter losses from precipitation and weathering during field storage are expected to reduce biomass quantity [38]. Storage losses of one percent per month were assumed [3]. Monthly shipments from the field are required to meet the biorefinery's monthly demand. A diesel fuel price of 680 \$ m^{-3} (the average price paid from 2005 to 2008 for bulk delivery) [39] was used in the computation of harvest and transportation costs.

Cost of transporting switchgrass biomass is based on the transportation cost equation developed by Wang[40]. Wang [40] estimated the cost of transporting biomass by assuming that a semi-tractor trailer will be used to transport switchgrass biomass bales from fields where bales are produced to the biorefinery. The capacity of the trailer was assumed to be 24 bales weighing approximately 0.9 Mg each. Therefore, a load would carry approximately 18 Mg dry biomass at moisture mass fraction ranging between 15% and 20%. Given the assumed diesel fuel price of 680 \$ m^{-3}, Wang's [40] cost estimation can be described as a function of the travel distance between the field and the processing facility. The equation is $TC_{ij} = 0.8799 + 0.1756 d_{ij}$, where TC_{ij} is \$ Mg^{-1} round trip cost of transporting 1.0 Mg of baled biomass and d_{ij} is the round-trip distance in kilometers.

BIOREFINERY COST

It was assumed that each biorefinery will produce only one output, ethanol. A gasification-fermentation technology was assumed [4] and [21]. Biorefinery capacity was fixed at 189 dam^3 y^{-1} of ethanol with an estimated capital cost of 379 M\$ and conversion rate of 0.35 m^3 of ethanol Mg^{-1} [4] and [21]. Operation and maintenance cost of the biorefinery was assumed to be 260 $ m^{-3}, and includes costs for labor, utilities, chemicals, taxes, repairs and insurance [4] and [21].

Model Equations

Following Tembo et al. [19] and others [21], [22], [23] and [24] the objective function of the model is to maximize the net present value (NPV) of the system:

$$\max \text{NPV}$$

$$Q_{jm}, A_{ilm}, XT_{ijkm}, XSIP_{ikm}, X_{ilm}, XP_{jkm},$$

$$XSI_{ikm}, XSIN_{ikm}, XSJ_{ikm}, \text{HUB}, \text{HUM} = \left(\left\{ \sum_{m=1}^{M} \sum_{j=1}^{J} \rho Q_{jm} - \sum_{i}^{I} \sum_{k}^{K} \delta_{k} \sum_{l=1}^{L} A_{ilm} - \sum_{i=1}^{I} \sum_{l=1}^{L} \zeta A_{ilm} - \sum_{i=1}^{I} \sum_{l=1}^{L} \alpha_{lm} A_{ilm} - \sum_{i=1}^{I} \sum_{l=1}^{L} \gamma_{lm} A_{ilm} \right.\right.$$

$$\left. - \sum_{i}^{I} \sum_{j}^{J} \sum_{k}^{K} \tau_{ij} XT_{ijkm} - \sum_{i=1}^{I} \sum_{k=1}^{K} \Gamma_{k} XSIP_{ikm} \right) - \sum_{j=1}^{J} \sum_{f=1}^{F} \text{POMC}_{f} \beta_{j} - \omega \text{HUM} - \omega \text{HUB} \right\} * \text{PVAF}$$

$$\left. - \sum_{j=1}^{J} \sum_{f=1}^{F} \text{AFC}_{f} \beta_{j} \right.$$

(1)

where the set M refers to months ($m = 1, ..., 12$ for Jan–Dec); J refers to potential biorefinery locations (j = Pontotoc, Washington, Canadian, Garfield, Okmulgee, Payne, Blaine, Carter, Grady, Kay, Woods, Comanche, Custer, Jackson, Texas and Woodward counties); I refers to potential biomass production regions (i = 77 Oklahoma counties); K refers to the switchgrass production system (k = established on cropland, established on improved pasture land); L refers to the type of land (l = cropland, improved pasture land); and F refers to facilities (f = Processing, Storage). Where ρ is the price of ethanol; Q_{jm} is the quantity of ethanol produced in month m by biorefinery at location j; δk is the cost of producing switchgrass with system k excluding cost of land, fertilizer and harvest; A_{ilm} is the land harvested in month m from land class l in county i; ζl is the cost of land class l; α_{lm} is the cost of applied nitrogen to land class l harvested in month m; γlm is the cost of applied P$_2$O$_5$ to land class l harvested in month m; τ_{ij} is the round-trip cost of transporting biomass from county i to biorefinery located at j; XT_{ijkm} is the quantity of biomass transported from county i in month m from system k to a biorefinery at location j; Γk is the cost of storing biomass in the field with production system k; $XSIP_{ikm}$ is the quantity of biomass placed in storage in month m from system k in county i; $\text{POMC}f$ is the cost of operating and maintaining type f facility; β_j is a binary variable for biorefinery at location j (1 if built, 0 otherwise); ω

is the annual cost of a mowing unit; HUM is an integer variable of the total number of mowing harvest units; ϖ is the annual cost of a raking-baling-stacking unit; *HUB* is an integer variable representing the total number of raking-baling-stacking harvest units PVAF; is the present value of annuity factor where PVAF $= (1 + r)T - 1/r(1 + r)T$; AFC_f is investment cost for facility type f in year 0 at location j.

The objective function in equation (1) is maximized subject to a set of constraints. Equation (2) is the cropland constraint equation that restricts total planted switchgrass area in a county on cropland to not exceed the quantity of available cropland ($POTACRE_{il}$) times a set proportion (BIPROP) of 10%.

$$\sum_{m=1}^{M} A_{ilm} - BIPROP * POTACRE_{il} \leq 0, \forall_i, = cropland \qquad (2)$$

Equation (3) imposes a similar restriction for improved pasture land BIPROP1 is the set proportion of 10% on improved pasture land.

$$\sum_{m=1}^{M} A_{ilm} - BIPROP * POTACRE_{il} \leq 0, \forall_i, = improved\ pasture\ land \qquad (3)$$

Equation (4) represents a yield balance used in computing the quantity of switchgrass biomass produced on the harvested lands.

$$\sum_{l=1}^{L} X_{ilm} - \sum_{l=1}^{L} A_{ilm} BYLD_{il} YAD_{km} \leq 0, \forall_i, k, m \qquad (4)$$

Where *Xilm* is the quantity of biomass harvested in month m from land class l in county i; BYLDil is the biomass yield from production in county i on land class l and YADkm is the biomass yield adjustment factor for production system k harvested in month m.

Equation (5) limits the months in which switchgrass can be harvested. YADkm is set to zero for the months of April, May and June, indicating no harvest in those months. Harvesting in those months in Oklahoma may damage switchgrass plants.

$$\sum_{l=1}^{L} A_{ilm} = 0 - \text{if } YAD_{km} = 0, \quad \forall_i, k, m \tag{5}$$

Equation (6) balances that the sum of biomass transported to the plant location from production regions at each source and in each month, plus biomass stored, with the sum of current production and the usable portion of stored biomass at the source county.

$$\sum_{l=1}^{L} X_{ilm} + \theta l_k XSI_{ikm-1} - \sum_{j=1}^{J} XT_{ijkm} - XSI_{ikm} = 0, \quad \forall_i, k, m \tag{6}$$

where θ_{lk} is the usable proportion of biomass from production system k stored in field (1 – storage loss %); XSI_{ikm} is the biomass stored in field in month m from system k in county i. Equation (7) balances the total biomass quantity transported to the plant plus the total storage loss with quantity harvested.

$$\sum_{i=1}^{i} \sum_{l=1}^{L} X_{ilm} - \sum_{j=1}^{J} \sum_{m=1}^{M} XT_{ijkm} - (1-\theta l_k) \sum_{m=1}^{M} XSI_{ikm} = 0, \quad \forall m \tag{7}$$

Equation (8) balances the total quantity of biomass harvested in addition to the quantity of biomass removed from field storage each month with the quantity of biomass transported from each county to the plant plus the amount of biomass placed in storage at the biorefinery.

$$\sum_{l=1}^{L} \sum_{l=1}^{L} X_{ilm} - \sum_{i=1}^{I} \sum_{j=1}^{J} \sum_{k=1}^{K} XT_{ijkm} + \sum_{i=1}^{I} \sum_{K=1}^{K} XSIN_{ikm} - \sum_{j=1}^{J} \sum_{K=1}^{K} XSIP_{ikm} = 0, \quad \forall_m \tag{8}$$

Where $XSINikm$ is the biomass removed from field storage in month m from system k and county i.

Equation (9) restricts plant processing capacity in each month at each location and equation (10) restricts monthly storage capacity at each biorefinery.

$$Q_{jm} - CAPP\beta_j \leq 0, \forall \quad j, m \tag{9}$$

Where CAPP is biorefinery processing capacity

$$\sum_{K=1}^{K} XSJ_{jkm} - CAP\beta_j \leq 0, \qquad \forall_{j,m} \qquad (10)$$

Where XSJ_{jkm} is the biomass stored in month m from system k onsite at biorefinery location j and CAP is the onsite biomass storage capacity at the biorefinery.

Equation (11) balances the quantity of biomass transported to the plant in month m minus the amount processed at the biorefinery in that month to the change in biomass storage inventory.

$$\sum_{i=1}^{i} XT_{ijkm} + \theta J_k XSJ_{jkm-1} - XSJ_{jkm} - XP_{jkm} = 0, \qquad \forall_{j,k,m} \qquad (11)$$

where θ_{jk} is the usable proportion of biomass from production system k stored onsite at the biorefinery (1 – storage loss %) and $XPjkm$ is the biomass processed in month m from system k by the biorefinery at location j.

Equation (12) balances the total biomass delivered from each production county to the biorefinery(ies) with the sum of processed biomass plus storage losses.

$$\sum_{i=1}^{i} \sum_{m=1}^{M} XT_{ijkm} - (1 - \theta J_k) \sum_{m=1}^{M} XSJ_{jkm} - \sum_{m=1}^{M} XP_{jkm} = 0, \qquad \forall_{j,k,m} \qquad (12)$$

· Equation (13) imposes a minimum biomass inventory (BINV) at the plant.

$$\sum_{K=1}^{K} XSJ_{jkm} - BINV\beta_j \geq 0, \qquad \forall j,m \qquad (13)$$

Equation (14) imposes ethanol production in each month to not exceed the capacity of the biorefinery(ies).

$$Q_{jm} - \sum_{K=1}^{K} \lambda_K XP_{jkm} \leq 0, \qquad \forall j,m \qquad (14)$$

Where λ_k is the quantity of ethanol produced from 1.0 Mg of biomass from production system k.

Equation (15) imposes a restriction on the number of endogenously determined mowing harvest units in any month to not exceed the available number of units.

$$\sum_{i=1}^{I} XHUM_{im} - HUM \leq 0, \qquad \forall_m \qquad (15)$$

Where $XHUM_{im}$ is the proportion of a mowing harvest unit used in month m in county i.

The sum of raking-baling-stacking harvest units used in each month is restricted by equation (16) to not exceed the total number of raking-baling-stacking harvest units endogenously determined by the model.

$$\sum_{i=1}^{I} XHUB_{im} - HUB \leq 0, \qquad \forall_m \qquad (16)$$

Where $XHUB_{im}$ is the proportion of a raking-baling-stacking harvest unit used in month m in county i.

Equations (17), (18), (19) and (20) ensure that the amount of harvested biomass in each month not exceed the harvesting capacity of the number of mowing harvest units and raking-baling-stacking harvest units.

$$CAPHUM_{im} = FWD_{im}DCAMHU_m \qquad \forall_{i,m} \qquad (17)$$

Where $CAPHUM_m$ is the capacity of a mowing harvest unit in month m; FWD_{im} is the field work days suitable for mowing in county i in month m; $DCAMHU_m$ is the daily capacity of a mowing harvest unit in month m.

$$\sum_{l=1}^{L} X_{ilm} - XHUM_{im}CAPHUM_{im} \leq 0, \quad \forall_{i,m} \tag{18}$$

Where $XHUM_{im}$ is the proportion of a mowing harvest unit used in month m in county i;

$$CAPHUB_{im} = BWD_{im}DCABHU_{m} \quad \forall_{i,m} \tag{19}$$

Where $CAPHUB_{m}$ is the capacity of a raking-baling-stacking harvest unit in month m; $BWDim$ is the number of field work days suitable for raking-baling-stacking in county i in month m; $DCABHU_{m}$ is the daily capacity of a raking-baling-stacking harvest unit in month m.

$$\sum_{i=1}^{L} X_{ilm} - XHUB_{im}CAPHUB_{im} \leq 0, \quad \forall_{i,m} \tag{20}$$

Equation (21) ensures that the raking-baling-stacking usage in each production region in each month does not exceed capacity.

$$XHUM_{im}CAPHUM_{im} - XHUB_{im}CAPHUB_{im} = 0, \quad \forall i,m \tag{21}$$

Equation (22) sets an upper bound on the number of biorefineries that can be built.

$$\sum_{j=1}^{J} \beta_{j} \leq b \tag{22}$$

Where b is the maximum number of biorefineries

Initially, the model was restricted to a single biorefinery ($b = 1$). The number of plants was incremented by one (the level of b was adjusted) and the model resolved until nearly 10% of the cropland and improved pasture land in the 77 counties was converted to switchgrass.

Equation (23) represents non-negative decision variables.

$$Q_{jm}, A_{ilm}, XT_{ijkm}, XP_{jkm}, XSI_{ikm}, XSIP_{ikm}, X_{ilm}, XSIN_{ikm}, XSJ_{jkm}, XHUM, XHUB \geq 0 \quad (23)$$

The quantity of mowing harvest units (HUM) and the quantity of cutting and raking-baling-stacking harvest units (HUB) are limited to be non-negative integer values.

The biorefinery location is restricted to be binary variable (equation (24)).

$$\beta j \in \{0,1\} \quad (24)$$

RESULTS

The estimated average (across all biorefineries) and marginal costs (the cost for each successive biorefinery) to deliver feedstock for one to nine biorefineries are reported in Table 2. Based on model assumptions and parameter values, the estimated cost to deliver 1.0 Mg of switchgrass feedstock is 54.91 $ for a single optimally located 189 dam^3 y^{-1} biorefinery. The 54.91 $ Mg^{-1} cost includes 13.33 $ (24% of the 54.91 $) for land rent, 15.64 $ (28%) for establish and maintenance, 0.45 $ (1%) for field storage, 15.04 $ (27%) for harvest and 10.45 $ (19%) for transporting the biomass from the field to the biorefinery. Cost to deliver feedstock increases as additional biorefineries are constructed. Our results indicate that on average there is a 49% (10.45 $ Mg^{-1} to 15.52 $ Mg^{-1}) increase in average feedstock transportation cost as the number of biorefineries increases from one to nine.

Table 2: Estimated average and marginal costs to deliver switchgrass biomass feedstock for one to nine biorefineries

Economic variable	Biorefineries (number)								
	1	2	3	4	5	6	7	8	9
Average cost across all biorefineries ($ Mg⁻¹)									
Land rent	13.33	14.17	13.62	13.62	14.17	14.05	14.56	15.23	15.88
Field storage	0.45	0.45	0.45	0.45	0.45	0.45	0.45	0.45	0.45
Establishment and maintenance	15.64	16.08	15.80	16.27	16.63	16.66	17.12	17.94	18.88
Harvest	15.04	14.54	14.58	14.69	14.32	14.64	14.56	14.34	14.06
Transportation cost	10.45	10.87	11.99	12.52	12.98	14.06	14.17	14.06	15.52
Total average cost	54.91	56.11	56.43	57.56	58.56	59.87	60.86	62.03	64.79
Marginal cost of each biorefinery ($ Mg⁻¹)									
Land rent	13.33	15.01	12.52	13.62	16.37	13.45	17.62	19.92	21.08
Establishment and maintenance	15.64	16.52	15.24	17.68	18.07	16.81	19.88	23.68	26.40
Harvest	15.04	14.04	14.66	15.02	12.84	16.24	14.08	12.80	11.82
Transportation cost	10.45	11.29	14.23	14.11	14.82	19.46	14.83	13.29	27.20
Marginal cost[a]	54.91	57.31	57.10	60.88	62.55	66.41	66.86	70.14	86.95
Increase in marginal cost relative to first biorefinery		4%	4%	11%	14%	21%	22%	28%	58%
Marginal biofuel feedstock component cost ($ m⁻³)									
Marginal cost of feedstock @ 0.35 m³ Mg⁻¹	157	164	163	174	179	190	191	200	248

ᵃThe cost of feedstock increases for each additional biorefinery. The average estimated cost for each unit of biomass delivered to the first biorefinery in the state is 54.91 $ Mg⁻¹. However, the average cost for each unit of biomass delivered to the ninth biorefinery in the state is 86.95 $ Mg⁻¹, 58% more.

The model optimally selects the least-cost location for each biorefinery. Fig. 2 and Fig. 3 illustrate optimal biorefinery location and counties that produce feedstock for a single and for nine biorefineries, respectively. As shown in Table 2 the cost to deliver switchgrass feedstock is greater for each successive biorefinery. Given the constraint that no more than 10% of the cropland and no more than 10% of the improved pasture land can be used to produce switchgrass, and based on the estimated yields, the state would be limited to nine 189 dam³ y⁻¹ capacity biorefineries. Insufficient biomass would be available to support a 10th biorefinery of that size. The cost to deliver biomass to the ninth biorefinery of 87 $ Mg⁻¹ is 58% greater than the cost to deliver to the first biorefinery (55 $ Mg⁻¹).

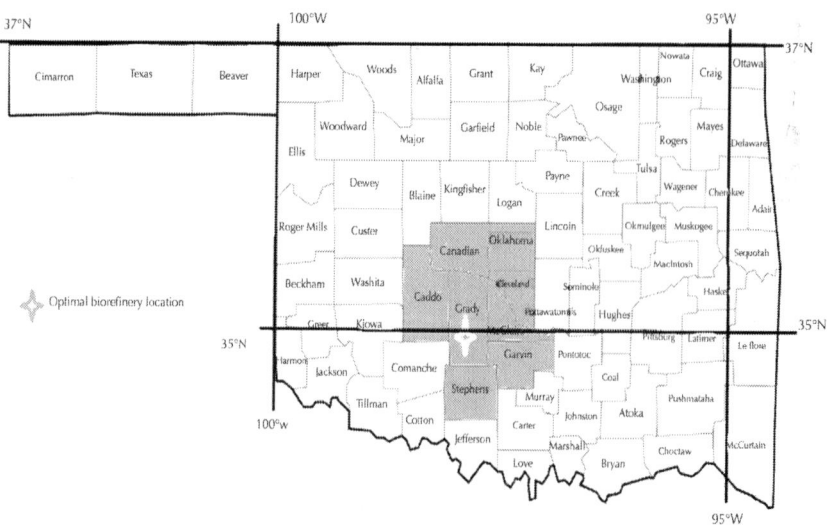

Figure 2: Optimal biorefinery location and counties (in gray) providing feedstock (from cropland and improved pasture land) for the least-cost biorefinery.

Figure 3: Optimal biorefinery locations and counties (in gray) providing feed-stock (from cropland and improved pasture land) for nine biorefineries.

Harvest cost is 21% lower for the ninth plant relative to the first plant. This finding follows from the assumption that harvest is conducted by coordinated harvest crews and from the assumption that mowing and rake-bale-stack harvest units are acquired in integer units and that the nine biorefineries would share harvest machines and harvest crews. One consequence is that excess harvest capacity is relatively less if the state has nine rather than only one biorefinery. This is not true across the entire spectrum of biorefineries. For example, harvest costs for the sixth plant are estimated to be 16.24 $ Mg^{-1}, 26% greater than the 12.84 $ Mg^{-1} harvest cost for the fifth biorefinery. The optimal number of harvest units, number of harvest machines and average investment in harvest machines for one to nine biorefineries are presented in Table 4. Moving from four to five biorefineries requires the addition of 10 rake-bale-stack units. However, moving from five to six biorefineries requires the addition of 13 rake-bale-stack units. The optimal number of purchased harvest machines, if shared across five biorefineries have almost no excess capacity. However, when the sixth biorefinery is added more harvest machines are required, but not all are used at full capacity.

For a single biorefinery the optimal number of mowing units (self-propelled windrower) is 15 and for nine biorefineries the optimal number of mowing units is 131. In addition, the optimal number of raking-baling-stacking units is 12 that includes 36 rakes, 36 balers, 12 field stackers, 36 tractors (41 kJ s⁻¹) for raking, and 36 tractors (149 kJ s⁻¹) for baling for a single biorefinery. For nine biorefineries the optimal number of raking-baling-stacking units is 100 that includes 300 rakes, 300 balers, 100 field stackers, 300 tractors (41 kJ s⁻¹) for raking, and 300 tractors (149 kJ s⁻¹) for baling.

Table 2 also includes an estimate of the feedstock cost per unit of ethanol produced. The estimated feedstock cost is 157 $ m⁻³ for the first biorefinery and 248 $, 58% greater for the ninth biorefinery. Clearly, the location and selection of land on which to establish switchgrass and the location of the biorefinery designed to use the feedstock will have economic consequences that will span the life of the plant and/or the life of the land use arrangements.

Table 2 also shows that for a single biorefinery the estimated storage cost is relatively low (0.45 $ Mg⁻¹). This estimate is a result of the assumption of a nine-month harvest window enabling harvested feedstock to be transported by truck soon after harvest in a just-in-time manner during nine months of the year. The nine month harvest window reduces the investment required in harvest machines and reduces the cost for feedstock storage relative to that of a system using a narrower harvest window. The average biomass storage cost per unit of feedstock is relatively unchanged as additional biorefineries are constructed.

The quantity of cropland and improved pasture land optimally leased, biomass harvested, and average yield for each of the nine biorefineries is reported in Table 3. To fulfill the requirement of a single biorefinery, a total of 539 km² are identified for use and assumed to produce 549 Gg annually. For a single biorefinery the weighted average switchgrass yield is 10.2 Mg ha⁻¹ but for ninth biorefinery the weighted average yield is 6.0 Mg ha⁻¹. Only 539 km² are optimally leased for the first biorefinery, but 917 km² are required for the ninth biorefinery.

Table 3: Cropland and improved pasture land leased, biomass harvested, and average yield for each of the nine biorefineries

BIOREFINERIES (NUMBER)	LAND HARVESTED (KM²)			BIOMASS HARVESTED (GG)			AVERAGE YIELD (MG HA⁻¹)		
	CROPLAND	IMPROVED PASTURE LAND	TOTAL	CROPLAND	IMPROVED PASTURE LAND	TOTAL^	CROPLAND	IMPROVED PASTURE LAND	WEIGHTED AVERAGE
1	382	156	539	424	125	549	11.1	8	10.2
2	521	41	562	517	32	550	9.9	7.9	9.8
3	327	195	522	387	163	551	11.8	8.4	10.5
4	247	375	622	259	290	549	10.5	7.7	8.8
5	559	59	617	510	40	550	9.1	6.9	8.9
6	294	295	590	329	220	549	11.2	7.5	9.3
7	550	139	689	459	90	549	8.3	6.5	8
8	560	252	812	432	118	550	7.7	4.7	6.8
9	467	450	917	302	251	552	6.5	5.6	6

^Quantity of biomass harvested differs across bio refineries as a result of differences in the number of harvest days across counties and subsequent storage losses.

Table 4: Optimal number of harvest units, number of harvest machines, and average investment in harvest machines for one to nine bio refineries[a]

Variable	Bio refineries (number)								
	1	2	3	4	5	6	7	8	9
Mowing units, self-propelled windrower, 142 kJ s^{-1}	15	30	43	58	72	88	103	117	131
Harvest units for raking-baling-stacking	12	23	35	47	57	70	81	91	100
Tractors (41 kJ s^{-1}) for rakes, rakes, tractors (149 kJ s^{-1}) for balers, balers (1.22 m × 1.22 × 2.44 m)	36	69	105	141	171	210	243	273	300
Bale transporter stackers	12	23	35	47	57	70	81	91	100
Total investment in harvest machines, M$	14.53	27.95	42.30	56.82	69.06	84.78	98.21	110.45	121.59
Marginal investment in harvest machines per biorefinery, M$	14.53	13.42	14.35	14.52	12.24	15.72	13.43	12.24	11.14

[a]These findings follow from the assumption that harvest is conducted by coordinated harvest crews and from the assumption that mowing and rake-bale-stack harvest units are acquired in integer units and that harvest machines and harvest crews could be shared across situations with multiple biorefineries. In most cases the "last" harvest unit "purchased" will not be used at full capacity.

As the number of biorefineries is increased, the model locates subsequent plants in regions with lower expected yields and relatively less dense cropland and improved pasture land. For a single biorefinery the average expected yield from cropland and improved pasture land is 11.1 Mg ha^{-1} and 8.0 Mg ha^{-1}, respectively, but for nine biorefineries the expected yields across all land converted are reduced to 6.5 Mg ha^{-1} and 5.6 Mg ha^{-1}. As the number of biorefineries is increased, subsequent plants are located in regions with less productive land requiring an increase in average feedstock transportation distances.

The breakeven ethanol price for each biorefinery, average breakeven ethanol price across plants, ethanol price equivalent to gasoline price, crude oil equivalent price and optimal plant locations with increased number of biorefineries are reported in Table 5. Investment and operating and maintenance costs for each biorefinery were assumed to be fixed at approximately 580 $ m^{-3}[4]. Differences in marginal breakeven ethanol price reported in Table 2 follow from the differences in the marginal costs of delivered feedstock. Thus, the breakeven price for the first biorefinery is estimated to be 740 $ m^{-3} (580 $ + 160 $). The breakeven price for the ninth biorefinery is estimated to be 830 $ m^{-3} (580 $ + 250 $). By this measure, the additional feedstock cost increases the cost to produce ethanol by 12%.

Table 5: Breakeven price of ethanol, gasoline equivalent price, price of crude oil and optimal location of the plants with increased number of biorefineries

Biorefinery (number)	Marginal breakeven price for next plant $ m^{-3}	Gasoline equivalent price $ m^{-3a}	Crude oil equivalent price $ Mg^{-1b}	Locations of the biorefineries
1	740	1130	1290	Grady
2	750	1140	1300	Grady & Garfield
3	750	1140	1300	Grady, Garfield, & Okmulgee
4	760	1150	1310	Grady, Garfield, Okmulgee, & Pontotoc
5	760	1160	1320	Grady, Garfield, Okmulgee, Pontotoc, & Woods
6	770	1180	1340	Grady, Garfield, Okmulgee, Pontotoc, Woods, & Washington
7	780	1180	1350	Canadian, Comanche, Garfield, Okmulgee, Pontotoc, Washington, Woodward
8	790	1190	1360	Blaine, Garfield, Grady, Jackson, Okmulgee, Pontotoc, Washington, Woodward
9	830	1270	1450	Blaine, Grady, Garfield, Jackson, Okmulgee, Pontotoc, Texas, Woods, Washington

[a]Gasoline equivalent price was estimated based on energy content of 21,100 MJ m^{-3} (lower heating value) of ethanol and 32,000 MJ m^{-3} of unleaded gasoline.

[b]Crude oil price calculated based on estimated value from a regression equation of the annual price of gasoline $ m^{-3} on the price of crude oil $ Mg^{-1} using 23 years (1986–2009) of historical data [41]. The estimated regression equation is: wholesale gasoline ($ m^{-3}) = 13.426 + 0.8652 × crude oil price ($ Mg^{-1}). By this measure for a crude oil price of 1286 $ Mg^{-1} the expected wholesale price of gasoline is 1128 $ m^{-3}.

Ethanol contains less energy (21,100 MJ m^{-3}) than unleaded gasoline (32,000 MJ m^{-3}). Energy contents are expressed here as Lower Heating Value (LHV). Based on the energy content (LHV) of ethanol relative to unleaded gasoline, 741 $ m^{-3} is equivalent to 1128 $ m^{-3} of gasoline, which is the expected wholesale price of gasoline if the price of crude oil is 1286 $ Mg^{-1}[42]. Given the assumptions that each biorefinery is the same size, has the same feedstock requirements and cost structure, as the number of the biorefineries is increased, the cost of delivered feedstock increases which increases the breakeven price of ethanol.

CONCLUSIONS

The cost to deliver feedstock increases as additional biorefineries are constructed. As additional biorefineries are located in regions with lower expected yields, more land is required, and as switchgrass must be established on more land, transportation distances increase. The breakeven price increases with the number of biorefineries as cost to deliver feedstock increases. Initial cellulosic biorefineries will have an opportunity for establishing a feedstock cost advantage by (a) carefully selecting a location for the biorefinery; (b) strategically selecting land for conversion to switchgrass; and (c) by acquiring long term land use rights for the selected land perhaps through negotiating long term leases.

DISCUSSION

In addition to the standard caveats associated with normative mathematical programming models, this study has several limitations and shortcomings. First, each biorefinery is assumed to be identical, operate at full capacity and have identical investment and operating and maintenance cost. Second, the model as executed was solved to first identify the optimal location for one biorefinery, then for two biorefineries, and eventually for nine biorefineries. These results would be appropriate for a single company that was seeking to determine how many biorefineries to build and where they should be located. Results may differ if the model was solved to identify the location of the next biorefinery subject to the added constraints that some biorefineries existed and the land leased by the existing biorefineries was not available for lease by the next facility. For example, the model as executed, when permitted to select the most optimal locations for a company that planned to construct seven biorefineries chose to locate them in Canadian, Comanche, Garfield, Okmulgee, Pontotoc, Washington and Woodward counties.

However, if the model had been forced to identify the location for the seventh biorefinery after the first six had been identified, plants would have already been located in Grady and Woods Counties. When adding the seventh plant under the assumption that no plants exist, the model chooses to add Canadian, Comanche, and Woodward counties and to not have plants in Grady and Woods counties. It is reasonable to assume that if plants already existed in Grady and Woods Counties, it would not be economical to close them and build new bio refineries in Canadian, Comanche, and Woodward counties. If the seventh plant was selected subject to the existence of the initial six plants the estimated marginal costs would be greater than those reported.

A third limitation is that land for conversion to switchgrass in each county was limited to 10% of the cropland and 10% of the improved pasture land. Local land leasing rate response to alternative levels of conversion from existing use to switchgrass remains to be determined. As more land is leased the expected lease rate will also increase resulting in an even greater expected feedstock cost for the nth biorefinery.

A fourth limitation is that switchgrass biomass yield variability was not considered. If yields are normally distributed and correlated within year across counties, total annual production from the land identified for leasing would be insufficient to meet the annual needs of the biorefinery in approximately half of the production years. Alternatively, biomass production in excess of biorefinery needs could be expected in approximately half of the years. As more information becomes available regarding switchgrass yield variability across years and across counties within years, additional modeling would be required to determine the economic tradeoffs among a strategy that includes year-to-year storage, a strategy that requires idling the plant during years of insufficient feedstock production, and a strategy to find and purchase other sources of feedstock.

Finally, the specific cost estimates follow from the model assumptions and parameter values. Adjustments to key parameter values such as biomass to ethanol conversion rate, fuel price, land lease rates, and other input prices would result in different estimates. However, the location and selection of land on which to establish switchgrass, the arrangements used to procure the feedstock, and the location of the biorefinery designed to use the feedstock, will have economic consequences that will span the life of the plant and/or the life of the land use arrangements.

ACKNOWLEDGMENTS

Funding for this project was provided by the USDA-NIFA, USDA-DOE Biomass Research and Development Initiative, Grant No. 2009-10006-06070, by the S amuel Roberts Noble Foundation, by the USDA National Institute of Food and Agriculture, Hatch grant number H-2824, and by the Oklahoma Agricultural Experiment Station. Support does not constitute an endorsement of the findings expressed.

REFERENCES

1.   de la Torre Ugarte DG, English BC, Jensen K. Sixty billion gallons by 2030: economic and agricultural impacts of ethanol and biodiesel expansion. Am J Agric Econ 2007;89(5):1290e5.

2.   United States Environmental Protection Agency. Renewable fuel standard program (RFS2) regulatory impact analysis. Washington,

DC: Assessment and Standards Division, Office of Transportation and Air Quality; 2010 Feb. Report No.:EPA- 420-R-10e006.

3.   Epplin FM, Clark CD, Roberts RK, Hwang S. Challenges to the development of a dedicated energy crop. Am J Agric Econ 2007;89(5):1296e302.

4.   Dutta A, Talmadge M, Hensley J, Worley M, Dudgeon D, Barton D, et al. Process design and economics for conversion of lignocellulosic biomass to ethanol: thermochemical pathway by indirect gasification and mixed alcohol synthesis. Golden, Colorado: National Renewable Energy Laboratory; 2011 May. p. 187, Report No.:TP-5100e51400.

5.   Larson JA, Yu TH, English BC, Mooney DF, Wang C. Cost evaluation of alternative switchgrass producing, harvesting, storing, and transporting systems and their logistics in the southeastern USA. Agric Financ Rev 2010;70(2):184e200.

6.   Timmons DS. Estimating a technically feasible switchgrass supply function: a western Massachusetts example. Bioenerg Res 2012;5(1):236e46.

7.   Khanna M, Dhungana B, Clifton-Brown J. Costs of producing miscanthus and switchgrass for bioenergy in Illinois. Biomass Bioenergy 2008;32(6):482e93.

8.   Duffy M. Estimated costs for production, storage and transportation of switchgrass. Iowa: Iowa State University Extension; 2007 October. p. 8, Report No.:PM 2042.

9.   Hess JR, Wright CT, Kenney KL. Cellulosic biomass feedstocks and logistics for ethanol production. Biofuels Bioprod Biorefin 2007;1(3):181e90.

10. Noon CE, Daly MJ. GIS-based biomass resource assessment with BRAVO. Biomass Bioenergy 1996;10(2e3):101e9.

11. Zhang F, Johnson DM, Sutherland JW. A GIS-based method for identifying the optimal location for a facility to convert forest biomass to biofuel. Biomass Bioenergy 2011;35(9):3951e61.

12. Graham RL, Liu W, Downing M, Noon CE, Daly M, Moore A. The effect of location and facility demand on the marginal cost of delivered wood chips from energy crops: a case study of the state of Tennessee. Biomass Bioenergy 1997;13(3):117e23.

13. Graham RL, English BC, Noon CE. A geographic information system-based modeling system for evaluating the cost of delivered energy crop feedstock. Biomass Bioenergy 2000;18(4):309e29.

14. Haque M, Biermacher JT, Kering M, Guretzky JA. Economics of alternative fertilizer supply systems for switchgrass produced in phosphorus-deficient soils for bioenergy feedstock. Bioenerg Res 2013;6(1):351e7.

15. Perrin R, Vogel K, Schmer M, Mitchell R. Farm-scale production cost of switchgrass for biomass. Bioenerg Res 2008;1(1):91e7.

16. Duffy M, Nanhou V. Costs of producing switchgrass for biomass in southern Iowa. In: Janick J, Whipkey A, editors. Reprinted from: Trends in new crops and new uses. Alexandria, VA: ASHS Press; 2002. pp. 267e75.

17. Brechbill SC, Tyner WE, Ileleji KE. The economics of biomass collection and transportation and its supply to Indiana cellulosic and electric utility facilities. Bioenerg Res 2011;4(2):141e52.

18. Vadas PA, Barnett KH, Undersander DJ. Economics and energy of ethanol production from alfalfa, corn, and switchgrass in the upper Midwest, USA. Bioenerg Res 2008;1(1):44e5.

19. Basnet A, Kenkel P. Feasibility of biomass harvesting cooperatives. Birmingham Alabama. In: Southern Agricultural Economics Association 2012 Annual Meeting; February 4e7, 2012. pp. 1e17.

20. Tembo G, Epplin FM, Huhnke RL. Integrative investment appraisal of a lignocellulosic biomass-to-ethanol industry. J Agric Res Econ 2003;28(3):611e33.

21. Kazi FK, Fortman JA, Anex RP, Hsu DD, Aden A, Dutta A, et al. Techno-economic comparison of process technologies for biochemical ethanol production from corn stover. Fuel 2010;89(1):S20e8.

22. Mapemba LD, Epplin FM, Taliaferro CM, Huhnke RL. Biorefinery feedstock production on conservation reserve program land. Rev Agric Econ 2007;29(2):227e46.

23. Mapemba LD, Epplin FM, Huhnke RL, Taliaferro CM. Herbaceous plant biomass harvest and delivery cost with harvest segmented by month and number of harvest machines endogenously determined. Biomass Bioenergy 2008;32(11):1016e27.

24. Wu J, Sperow M, Wang J. Economic feasibility of a woody biomass-based ethanol plant in central Appalachia. J Agric Res Econ 2010;35(3):522e44.

25. Haque M, Epplin FM. Cost to produce switchgrass and cost to produce ethanol from switchgrass for several levels of biorefinery investment cost and biomass to ethanol conversion rates. Biomass Bioenergy 2012;46:517e30.

26. Weyerhaeuser. Low-cost, high value harvesting. Federal Way, WA: Weyerhaeuser Company [Internet]; 2012 [cited 2012 Nov 12]; [about 1 screen]. Available from: http://www. weyerhaeuser. com/Businesses/Timberlands/ OptimizingHarvests.

27. Barros O, Weintraub A. Planning for a vertically integrated forest industry. Operat Res 1982;30(6):1168e82.

28. Jones PC, Ohlmann JW. Long-range timber supply planning for a vertically integrated paper mill. Eur J Oper Res 2008;191(2):558e71.

29. United States Department of Agriculture. United States summary and state data. 2002 census of agriculture. Washington DC: Research, Education, and Economics, National Agricultural Statistics Service; 2004 June. Report No.: AC-02-A-51. Geographic area Series 2004;1(Pt 51).

30. National Agricultural Statistics Service (NASS). Land values and cash rents 2010 summary. Washington, DC: United 318 biomass and bioenergy 66 (2014) 308 e319States Department of Agriculture; 2010 Aug. Report No.: ISSN: 1949-1867.

31. Farm Service Agency [internet]. [May 2013]-. Conservation Reserve Program. Monthly Summary. Washington DC: United States Department of Agriculture; May, 2012 [cited 2013 July 13]; [about 25 screens]. Available from: http://www. fsa.usda. gov/Internet/FSA_File/june2013crpstat.pdf.

32. Basnet A, Depona T, Hedges W, Dicks MR. Potential biomass yields in the south central US. Corpus Christi, Texas. In: Southern Agricultural Economics Association 2011 Annual Meeting; February 5e8, 2011. pp. 1e24.

33. Haque M, Epplin FM, Taliaferro CM. Nitrogen and harvest frequency effect on yield and cost of four perennial grasses. Agron J 2009;101(6):1463e9.

34. Epplin FM, Haque M. Conversion of millions of acres to the production of biofuel feedstock. J Agric Appl Econ 2011;43(2):385e98.

35. [35] Vogel KP, Brejda JJ, Walters DT, Buxton DR. Switchgrass biomass production in the Midwest USA: harvest and nitrogen management. Agron J 2002;94(3):413e20.

36. Hwang S, Epplin FM, Lee B, Huhnke RL. A probabilistic estimate of the frequency of mowing and baling days available in Oklahoma USA for the harvest of switchgrass for use in biorefineries. Biomass Bioenergy 2009;33(8):1037e45.

37. Thorsell SR, Epplin FM, Huhnke RL, Taliaferro CM. Economics of a coordinated biorefinery feedstock harvest system: lignocellulosic biomass harvest cost. Biomass Bioenergy 2004;22(4):327e37.

38. Cundiff JS, Grisso RD. Containerized handling to minimize hauling cost of herbaceous biomass. Biomass Bioenergy 2008;32(4):308e13.

39. National Agricultural Statistics Service (NASS). United States Department of Agriculture. Agricultural prices 2008 summary. Washington, DC: Agricultural Statistics Board; 2009 Aug. Report No.: Pr 1e3(09).

40. Wang C. Economic analysis of delivering switchgrass to a biorefinery from both the farmer's and processor's perspectives [Master's thesis]. Knoxville, Tennessee: The University of Tennessee; 2009.

41. U.S. Energy Information Administration [Internet]. Annual Cushing, OK WTI spot Price FOB. Washington DC: Petroleum and Other Liquids; 2010 [cited 2013 Jan 15]. Available from: http://tonto.eia.doe.gov/dnav/pet/hist/LeafHandler.ashx?n¼pet&s¼rwtc&f¼a.

42. Bioenergy.ornl.gov [Internet]. Oak Ridge, TN: Oak Ridge National Laboratory. Bioenergy feedstock development Programs; [year unknown]; Convers factor [cited 2014 Jan 8]; Available from: https://bioenergy.ornl.gov/papers/misc/ energy_conv.html.

Indicators of Bioenergy-Related Certification Schemes – An Analysis of the Quality And Comprehensiveness for Assessing Local/Regional Environmental Impacts

Markus A. Meyer and Joerg A. Priess

Helmholtz Centre for Environmental Research – UFZ, Department of Computational Landscape Ecology, Permoserstr. 15, 04318 Leipzig, Germany

ABSTRACT

Bioenergy is receiving increasing attention because it may reduce greenhouse gas emissions, secure and diversify energy supplies and stimulate rural development. The environmental sustainability of bioenergy production systems is often determined through life-cycle assessments that focus on global environmental effects, such as the emission of greenhouse gases or air pollutants. Local/regional environmental impacts, e.g., the impacts on soil or on biodiversity, require site-specific and flexible options for the assessment of environmental sustainability, such as the criteria and indicators used in bioenergy certification schemes.

In this study, we compared certification schemes and assessed the indicator quality through the environmental impact categories, using a standardized rating scale to evaluate the indicators. Current certification schemes have limitations in their representation of the environmental systems affected by feedstock production. For example, these schemes predominantly use feasible causal indicators, instead of more reliable but less feasible effect indicators. Furthermore, the comprehensiveness of the depicted environmental systems and the causal links between human land use activities and biophysical processes in these systems have been assessed. Bioenergy certification schemes seem to demonstrate compliance with underlying legislation, such as the EU Renewable Energy Directive, rather than ensure environmental sustainability. Beyond, certification schemes often lack a methodology or thresholds for sustainable biomass use. Lacking thresholds, imprecise causal links and incomplete indicator sets may hamper comparisons of the environmental performances of different feed stocks. To enhance existing certification schemes, we propose combining the strengths of several certification schemes with research-based indicators, to increase the reliability of environmental assessments.

INTRODUCTION

Bioenergy is receiving increasing attention because it is assumed to be associated with the following major advantages over fossil fuels [1], [2], [3] and [4]:

- Reduction of greenhouse gas (GHG) emissions and strengthening of the environmental sustainability of energy provision
- Securing and diversifying the energy supply
- Positive socioeconomic impacts such as increased energy access in developing and jobs in developed countries

The arguments in favor of bioenergy can be summarized under the concept of sustainability as defined by the Brundtland Commission [5]. The aspects listed above show that several dimensions of sustainability are of importance, namely the economic, environmental and social dimensions [6]. According to neoclassical theory, economic sustainability is ensured through market mechanisms [7]. Environmental and social sustainability are often not ensured through these mechanisms and require government interventions, for example, quotas for bioenergy or subsidies to overcome market failures [8]. Even if environmental and social sustainability are considered for bioenergy, Robbins [9] stated that it is currently unclear how to assess the sustainability of bioenergy from both environmental and socioeconomic perspectives.

The major environmental impact categories of bioenergy feedstock production have been summarized to GHG emissions, air pollutants, soil quality, water quality, water availability or quantity, biodiversity and land-use and land-use change (LU/LUC) based on scientific literature [10], [11], [12] and [13] and broader stakeholder panels [14]. To a great extent, the environmental sustainability of bioenergy production systems is evaluated with well-established life-cycle assessments (LCAs), assessing large-scale or globally occurring environmental effects, such as GHG emissions or air pollutants, along the major steps of the supply chain [10] and [15]. The highly site-specific and locally/regionally occurring environmental impacts of feedstock production in the first step of most of the bioenergy supply chains are difficult to assess in LCAs. Impacts on soil quality, biodiversity and land use change, water availability and water quality [16] and [17] are often insufficiently covered. These limitations comprise necessary but missing regional thresholds to ensure the stability of the ecological system. Such thresholds are not easily integrated into highly standardized LCAs. Existing LCAs assessing environmental impacts often disregard the interaction for example between different regulating ecosystems services (ESS) and biodiversity, such as the buffering capacity of environmental impacts of agriculture or forestry [18] and [19]. In the context of bioenergy feed stocks and sustainability, this type of assessment of interactions is supposed to extend the EU RED, i.e., the provision of "basic ecosystem services" such as erosion control should be accounted for if biomass is produced for bioenergy [20]. Dale et al. [21] recommend to determine water quality and soil quality impacts of bioenergy feedstock production in addition to LCAs, e.g., nutrient export to water bodies or soil loss. A regional water quality assessment will more likely allow to determine, whether regional thresholds of nutrient exports that ensure good ecological status of water bodies are met.

Site-specific and flexible options for the assessment of local/ regional environmental impacts and other aspects of sustainability could be sets of criteria and indicators (C&Is) as used in certification schemes. Such a site-dependent audit approach allows assessing the environmental impacts and their interactions mentioned above. C&Is are currently under development or are at an early stage of implementation for bioenergy but have been extensively applied for a longer period to other products from forestry or agriculture. Examples of C&is are the Forest Stewardship Council (FSC) for timber or the Sustainable Agriculture Network (SAN) as a label for Good Agricultural Practices [2]. Especially FSC provides nationally or regionally adapted indicator sets [22]. Several bioenergy certification schemes are used to demonstrate compliance with the EU Renewable Energy Directive 2009/28/EC (EU RED) [23].

Despite the common aim of EU RED compliance for most of the bioenergy schemes, an increasing number of alternative schemes may contribute to confuse stakeholders and decrease the acceptance of certification schemes in general [24] and [12]. On the one hand, comprehensive and clearly defined requirements may exclude producer groups [2], e.g., in developing countries, and augment certification costs due to increasing effort, such as audits. On the other hand, vaguely defined and less comprehensive schemes may allow for a higher market penetration, but more likely disregard major environmental or social impacts and are not acknowledged by NGOs [25] and [26]. An increase in EU imports of biomass for bioenergy might induce or enhance deforestation in countries with prevailing primary forests [27] and the need to export goods. Thus, overexploitation is more likely to occur in developing countries than in developed countries. To avoid or abate e.g., deforestation, a set of C&Is must be agreed upon internationally to cover international biomass trade [28]. International criteria might exceed the local requirements for bioenergy sustainability or might set foci other than the locally intended ones [29]; e.g., criteria might focus on environmental aspects in developed countries, such as sequestering carbon or halting biodiversity loss instead of ensuring food security in developing countries [13]. Such potential discrepancies may provide additional obstacles for implementation.

Beyond existing reviews [2], [12], [13], [26] and [29], this paper, assesses the comprehensiveness and quality of indicators used by bioenergy, forestry and agricultural certification schemes. Against the background of conflicting goals for bioenergy certification discussed above, we develop and apply standardized rating scales for indicators grouped into six environmental impact categories to identify their reliability and feasibility. We focus on local/regional environmental impacts, which require site-specific information, affect predominately the local/regional environment and are usually not covered by LCAs. Beyond rating the individual indicators, certification schemes are evaluated at the scheme level based on the ESS cascade [30] to analyze their comprehensiveness and the quality of the representation of the potentially affected environmental system. The aim is to test whether certification schemes are able to show trade-offs between biomass use and other ecosystem services.

MATERIAL AND METHODS

Selection of Certification Schemes and Indicator Sets

In this paper, indicator sets for certification have been selected for evaluation. We used sets from bioenergy, agriculture and forestry. The latter two have the advantage of a much longer lasting application of C&Is. Concentrating on the currently rather limited number of specific schemes for bioenergy would have led to a very small set of C&Is, ignoring relevant and important C&Is applied in related sectors.

First, the EU might consider the extension of bioenergy specific with forestry schemes as a relevant policy option for solid biomass for bioenergy in the EU, e.g., by using additional forestry indicators for sustainability certification [31]. Therefore, an evaluation of studies is conducted, assessing the environmental impacts of forest management with a focus on bioenergy production. To identify major characteristics of forestry certification schemes, we selected the FSC and the Sustainable Forestry Initiative (SFI), a major scheme of the meta-standard "Programmed for the Endorsement of Forest Certification" (PEFC), which are globally dominating and largely applied certification schemes in forestry [2] and [32]. We avoided meta-standards since they typically do not have indicators sets for the actual environmental assessment.

Secondly, new technologies to enhance the transport, storage and co-firing characteristics, such as Torre faction, are under development. These technologies might create additional feedstock options, for instance agricultural residues, such as straw, shells and others, which currently may be used to a limited extent [33]. Therefore, overarching and globally applied agricultural certification schemes, i.e., SAN and Global Good Agricultural Practice (Global GAP), are needed to cover feed stocks not targeted by bioenergy certification schemes, predominately aiming at selected bioenergy crops. The relevance of agricultural certification schemes shows NTA 8080 and other bioenergy certification schemes as they use agricultural certification schemes, which we also selected in this paper, to ensure compliance with environmental sustainability requirements [13]. Despite the fact that GBEP is no operational certification scheme, we included it in our assessment since its indicator set reflects the consensus of numerous governments and international institutions and because it is a framework to assess bioenergy sustainability [12].

Requirements and Rating Scales for Indicator Evaluation

The major requirements for indicators are reliability and conceptual soundness, feasibility, i.e., measurability and practicality, and relevance for the end user [2], [34], [35] and [36]. The requirements for an indicator discussed in this section are rated on a five step scale. Bock staller et al. [34] have demonstrated the methodological suitability of such an approach at the indicator level by evaluating sets of agree-environmental indicators for crop production and farming systems, which are methodologically comparable to the certification scheme indicators evaluated in this paper.

We rate the individual indicators for feasibility in three requirement subcategories and for reliability in four requirement subcategories, two exemplary requirement subcategories each are listed in Table 1 and the remaining ones in Appendix A.

Table 1: (Upper part) Rating scale for the reliability of indicators, subcategory Indicator type adapted from Bock staller et al. [34]; (lower part) rating scale for the feasibility of indicators, subcategory required resources (assessment interval)

Indicator type (cause vs. effect-related)	
1 Driver	Management practice
2 Driver	Management practices related to state or impact
3 Pressure	Release of pollutants or sediment
4 State	Concentration of pollutant in environmental compartment
5 Impact	Environmental changes attributable to pollutants or sediments

Required resources (assessment interval)
1 Daily assessment/measurements required
2 Seasonal assessment/measurements required
3 Annual assessment/measurements required
4 Less than annual measurements
5 No measurement, only completing a survey

The first rated subcategory for reliability is the Indicator type [34] and [37]. For practical implementation, we followed the logic of the **D**riving forces – **P**ressures – **S**tates – **I**mpacts – **R**esponses (DPSIR) framework of the European Environment Agency [36], extending preceding frameworks, such as the Pressure–State–**R**esponse framework, applied by the OECD and the UN [38]. We present an application example for the DPSIR framework for rising wood pellet demand, conceptually based on Bock staller et al. [39] and Svarstad et al. [38]. A rising demand of wood pellets may require to apply more fertilizer for shorter rotation cycles of forest plantations, e.g., Pinups spp., (Driving force). Consequently, increased fertilizer application may increase the nutrient runoff to surface water bodies (Pressure), which may lead to higher nutrient concentrations (State), i.e., possibly eutrophication, which may change e.g., the species composition (Response). Thus, an indicator of an environmental pressure such as the nutrient load from pine plantations on a water body would be rated as "three" on the five step scale, and a state indicator such as the nutrient concentration in a river would be rated as "four" or the nutrient application rate in the driver category as "one". The closer the assessment is to the environmental impact, the more information on the environmental impact is expected to be considered. The second subcategory for reliability is the Validity of indicators. We rate the validity, according to a rating scale, see Table A.1 in Appendix A, modified from Bockstaller et al. [34], which has been developed by Bock staller and Girard in [40]. We rate the indicators (i) based on scientific literature, i.e., whether peer-reviewed articles use and confirm the exact indicator (value 4), whether the indicator is under debate in the scientific literature (value 3), only confirm the calculation method of the indicator or even reject the indicator (value 2). (ii) Other options are that the indicator needs to agree with locally collected data (value 5) or is typically gained from a validated model (value 4), a partly or only regionally validated model (value 2). If no validation is possible due to the rating in the subcategory Indicator type rated as given for indicators on management practices (value 1 or 2), we rate the indicator with a value of "three". The third subcategory for reliability is the Response time since an immediate response or a response at least in the time frame of political decision making [10] and [36] enable timely detection and counteraction to the expected or observed environmental problems. We rate the response time of indicators based on peer-reviewed publications.

The first subcategory for feasibility is the Data requirement, assessing the ease of data access [2], [34], [36] and [39]. We rate indicators based on (i) the nature of the data, i.e., whether it can be obtained from authorities or other data sources (value 5), requires questioning the feedstock producer (value 4) or measurements are required (value 1–3). (ii) The measurement scale is additionally used for the rating [41], i.e., whether indicator data has to be measured at each field or farm individually (value 1) or whether one regional assessment is sufficient for the indicator (value 3). In addition, indicators may be attributed to the field/farm or the regional scale depending on the individual case (value 2), e.g., influenced by farm size (group certification) or an imprecise definition of the indicator in the certification scheme. The second subcategory for feasibility is the Qualification requirement [39], [34] and [2] covering the ease or difficulty to assess an indicator due to its specificity or the required expert knowledge (requirements defined in Appendix A). High qualification requirements may be an obstacle for small scale producers, especially in developing countries [24]. The third subcategory for feasibility is the required resources (assessment interval), i.e., the frequency of possible measurements influences the effort and costs for certification. The fourth subcategory for feasibility is clearly defined thresholds. We rate the existence of target values, reference conditions or thresholds because their availability influences the measurability [11]. A threshold or a possible source to derive it provided by the scheme facilitates the interpretation of feedstock impacts regarding sustainability during the auditing process [41].

The relevance of an indicator first depends on its acceptance by stakeholders, i.e., whether the indicator is suitable to address a certain environmental impact category [36], and secondly on the degree to which stakeholders are involved in the selection process [26]. Data on the preferences of stakeholders is only available for criteria or for the even higher aggregation level of environmental impact categories, but is not available for the corresponding indicators (c.f. Buchholz et al. [35]). The lack of data might also be due to the fact that the development and choice of the rather technical indicators are related to the expertise of the practitioners or scientists. Therefore, the relevance of the indicators cannot be rated but will be checked indirectly by its fit to the relevant environmental impact categories.

We rate indicators that provide direct information about the occurrence or avoidance of environmental impacts. The indicators are aggregated by local/regional environmental impact category on a composite scale. In this context, a composite scale is the combination of several indicators into a thematic category, i.e., we compute the arithmetic mean of all indicators per certification scheme per environmental impact category and the indicator subcategories respectively. Similarly, the standard error of the mean (SEM) is calculated to assess the uncertainty of the arithmetic mean. We assess the indicator sets for the environmental impact categories soil quality, water quality, water availability or quantity, biodiversity and LU/LUC. Soil quality indicators cover indicators on both the management of soils and soil properties. Water quality and availability indicators assess both management activities with an impact on water bodies as well as state indicators of water bodies. Biodiversity indicators may assess the state of conservation areas, species composition or management activities for biodiversity. LU/LUC indicators give information on characteristics of a land use, e.g., carbon payback time, or assess whether no-go areas according to the EU RED definition have been converted for bioenergy feed stocks. The composite scale other comprises indicators without a link to the listed environmental impact categories, which are related to the environmental stability of a system such as indicators on sustainable harvest levels. If applicable, indicators are attributed to two composite scales if a clear link to both is given, e.g., "no conversion of areas of high conservation value" to biodiversity and LU/LUC or "no removal of coarse woody debris" to soil quality and biodiversity.

Internal consistency is ensured by excluding indicators that do not directly measure environmental impacts, i.e., contextual knowledge is used according to Coste et al. [42]. Background knowledge on the environmental indicators, e.g., given by the certification scheme, allows to categorize the indicators. Internal consistency is required since the arithmetic mean should only be calculated for indicators that measure the same latent variable, i.e., environmental impact category. We exclude indicators, for example, if they assess whether legislation is covering environmental impacts, e.g., on water quality. In this case, certification schemes assume that environmental impacts are avoided (complying with existing regulations).

We list the indicators we included and excluded for each scheme in Table 2.

Table 2: Number of indicators analyzed for each scheme and each environmental impact category (=composite scale) and abundance of aspects in certification schemes excluded from evaluation to ensure internal consistency of composite scales; these results are based on CSBP [43], GBEP Task Force [14], GGL [44], GlobalGAP [45], ISCC [46], IWPB [47], Netherlands Standardization Institute [48] and [49], REDcert [50], RSB [51], SAN [52] and forestry [29], [32], [53], [54], [55] and [56]. For GGL, the agricultural source criteria (GGL2) are assessed

Composite scales		GBEP	NTA8080	ISCC	REDcert	GGL	RSB	CSBP	IWPB	SAN	GlobalGAP	Forestry
Total	87											
Soil quality	31	1	9	11	2	0	5	5	8	5	2	30
Water quality	17	2	4	6	7	4	3	6	3	4	7	6
Water availability	9	3	2	2	0	1	1	2	3	1	2	6
Biodiversity	18	3	5	1	1	1	7	3	1	10	2	13
LU/LUC	9	5	3	2	1	0	1	0	3	1	1	1
Others	3	2	0	0	0	0	0	0	1	0	0	3
Abundance of excluded aspects												
Off-site handling rules and machinery maintenance (e.g., disposal of plant protection product containers)	46	1	1	21	7	1	0	0	0	3	12	0

Demonstration of compliance with existing legislation or other rules such as certification schemes, manuals or rules (e.g., registration of product use)	32	0	3	5	3	0	4	3	6	5	3	0
Management plan or other unspecified action or goal required	33	0	0	2	1	2	19	4	0	2	3	0
Qualification and training of staff	10	0	0	2	1	0	0	1	3	1	2	0
Generic monitoring (e.g., soil quality has to be assessed)	6	0	0	0	0	2	2	0	0	2	0	0

The Ecosystem Service Cascade for Evaluation of Certification Schemes

Assessing certification schemes by only looking at indicators individually would disregard the schemes' quality and comprehensiveness concerning the use of environmental systems and the services/ disservices derived thereof. A widely accepted concept to determine and quantify the human use of the environment is ESS [57] and [58].

The ESS cascade [30] is a conceptual framework used to connect ESS to the underlying ecosystem structures and processes and to the human benefits derived from the use of the ecosystem. Ecosystem structures and processes are the basis to derive thresholds for the sustainable provision of an ESS [30] and [57], i.e., the ecosystem capacity. For example, the ecosystem capacity can be used to answer questions about the critical limits or thresholds [59] for e.g., the extraction of tree biomass to sustain forest stocks. Because this evaluation focuses on local/regional environmental impacts, it is beyond our scope to depict the socioeconomic components of the ESS cascade, i.e., the human benefits and (monetary) values. We focus on biophysical and ecological structures and functions and their alteration due to the use of ESS. The ecological and the socioeconomic systems are linked by the use of ESS [60], e.g., biomass use. In practice, the ESS cascade has been used as a conceptual framework to embed indicators of different provisioning services, e.g., biomass production [61] and [62], and regulating services, e.g., water purification [63], of the underlying environmental systems. In addition, the ESS cascade has also been used to visualize the interaction of indicators within and between the different components of the ESS cascade [62] and [64]. Maes et al. [63] and Van Oudenhoven et al. [62] add land management to the beforehand mentioned components of the ESS cascade. The necessity of including land management was previously stated by Haines-Young and Pots chin [30] but was not implemented. Like Ojima et al. [65], we included land management aspects because indicators of ESS describe the use of natural capital but do not provide insight into the extent that the use of ESS is altered by human land use activities, i.e., agricultural practices such as irrigation or fertilization or conservation measures such as field margins for biodiversity.

In this study, we use the term "human land use activity" because this term includes land management, land conversion and changes in the structure of the landscape [66]. Therefore, indicators of human land use activities enable the assessment of the intensity of land use associated with different types of and options for biomass provision. For example, changes in production practices or landscape planning are likely to affect ecosystems, i.e., the structures, processes and capacity. A better representation of the interaction of human land use activities, ecosystems and ESS use might help to identify environmentally especially harmful biomass use and land management practices. More reliable results could allow decision makers to better target, e.g., mitigation activities.

In this study, the ESS cascade is extended from a conceptual to an analytical framework for bioenergy feedstock production (Fig. 1). The ESS cascade is converted and expanded into an analytical tool to assess the quality of certification schemes. The latter are implemented within the framework to assess the sustainability of feedstock provision with environmental C&Is; i.e., the adverse environmental impacts should be revealed to facilitate mitigation or avoidance as requested by Van Dam et al. [13]. Thus, the extended ESS cascade is applied to investigate whether certification schemes represent biophysical processes for feedstock production in a qualitatively and quantitatively useful manner. We apply the widely used "Common International Classification of ESS – CICES" v4.3 [67], which has undergone several rounds of international review and consultation, to ensure assessing all major ESS, which may be affected by bioenergy feedstock production.

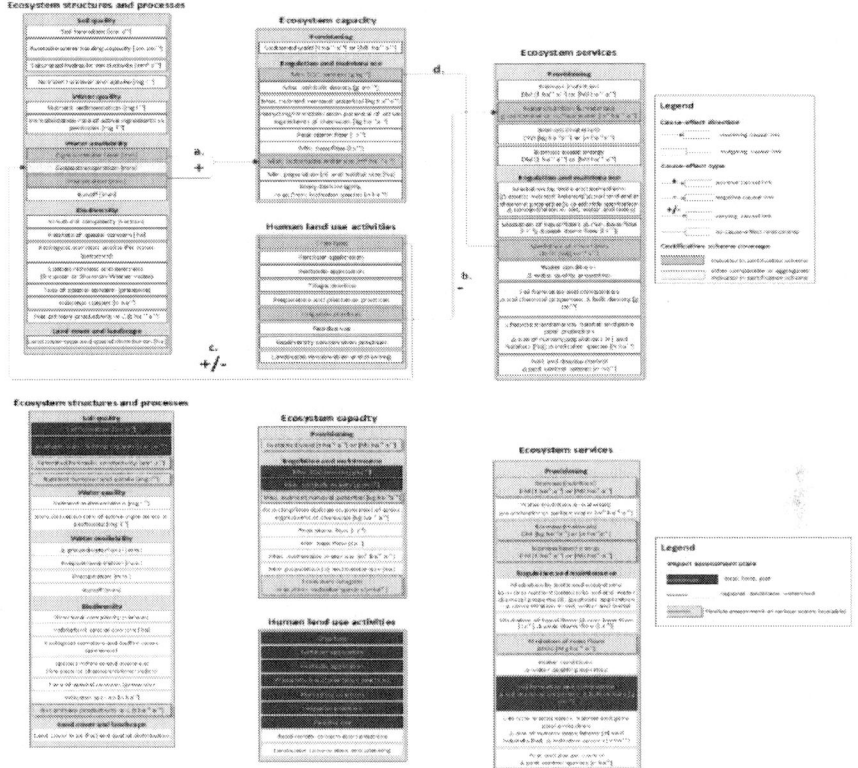

Figure. 1: (upper part) ESS cascade (modified from CICES [67], Maes et al. [63], Potschin and Haines-Young [60], Van Oudenhoven et al. [62]) as an analytical framework to evaluate certification schemes for bioenergy feedstock production; the components shown are ecosystem structures and processes (underlying biophysical mechanisms), ecosystem capacity (sustainability thresholds for ESS use) and ESS (actual use of ESS or creation of disservices). The arrows indicate a. positive, b. negative, c. varying and d. no causal link. The selected indicators are adapted to the major impacts of bioenergy production identified from Dale and Beyeler [68], De Groot et al. [57], Haines-Young and Potschin [30], Kandziora et al. [64], Kienast et al. [69], Lattimore et al. [53], McBride et al. [11], McElhinny et al. [70], Schoenholtz et al. [55], Wascher [71]. (Lower part) Spatial impact assessment scales of the ESS cascade adapted for bioenergy feedstock production. The impact assessment scales are generally based on De Groot et al. [57] and Efroymson et al. [10] and are specifically based on Sposito [72] for hydrology and Turner et al. [73] for landscape patterns.

The mapping used for the certification scheme indicators is presented in Fig. 1. For the different certification schemes we analyzed, we focused especially on the representation of causal links and the coverage of ecosystem structures and functions represented in the extended ESS cascade, i.e., the quality of the representation of the environmental system. For example, does a certification scheme include indicators that would reveal if biomass use affected other ecosystem services such as surface or groundwater provision? Does a certification scheme include the link from fertilized pine plantations to a possible ground- or surface-water pollution and does it provide the relevant indicators on, e.g., water quality and fertilization practices? We took the individual indicators per certification scheme, related them to the environmental system and indicated the causal links and components covered.

For an overview, we counted the actual number of indicators for each of the four components of the ESS cascade displayed in Fig. 1 and rated them on a three step scale based on thirds. For causal links, the certification schemes are compared with their peers. The certification scheme with the highest number of causal links has the best rating, i.e., 100%, and is used as a benchmark and rated as done for the indicators. The indicators and causal links for each scheme are displayed in Appendix A.

The following three types of common causal links and links without cause–effect relationships are found in the evaluated certification schemes and indicator sets:

a. **Positive causal link (Increase in X causes an increase in Y):** **Example -** "The participating operator provides objective evidence demonstrating that her/his/its biomass/biofuels operation(s) does/do not contribute to exceeding the replenishment capacity of the water table(s) [...]," RSB[51]. This statement implies that the maximal sustainable water use does not negatively affect the groundwater table and is adapted to the local level of precipitation. Therefore, both a higher precipitation and a higher change of the groundwater table, i.e., a lower decline, may result in a higher maximal sustainable water use.

b. **Negative causal link (Increase in X causes a decrease in Y):** **Example-** The feedstock provider measures the water use per area and uses irrigation techniques that conserve water most, e.g.,

CSBP [43]. In other words, if more irrigation techniques with low water use are applied (replacing inefficient technologies), the use of water units per unit bioenergy feedstock will decrease per ha.

c. **Varying causal link (Increase in X causes an increase or decrease in Y): Example** - "Have systematic methods of prediction been used to calculate the water requirement of the crop?" Global GAP [45]. Options for actions are suggested in the explanation of the indicator. The actions may be operationalized as follows: The amount of water used varies with the crop type. Hydrologic ally, the upward flux of water via plants and soil is termed evapotranspiration. The choice of a crop may increase or decrease evapotranspiration. Because this biophysical flux is not named in the indicator, but is only implicitly considered, it is highlighted in yellow.

d. **No cause–effect relationship**: The soil organic carbon content is maintained or improved, e.g., GBEP Task Force [14]. The definition of the indicator specifies both the ecosystem capacity and the parameter to be measured to determine the ESS use, i.e., mediation of mass flows. Here, a thematic link between ecosystem capacity and ESS is given instead of a cause–effect relationship.

Additionally, we need to assess how certification schemes are able to overcome the challenge of the necessity of assessing (i) environmental impacts at scales beyond the field/farm level [12] and (ii) the interaction and accumulation of environmental impacts beyond different spatial scales [10] and [37] and how to distribute target values or thresholds [74] and [75]. Within this study, the relevant spatial scales from both the literature on actual indicators and from specific studies on scales to determine specific environmental parameters are shown in Fig. 1. Because this study focusses on local/regional scale environmental impacts, there are no indicators included beyond those scales. Local scale, also plot or field scale, is typically areas less than 1 km² and regional, also landscape or watershed scale ranges from 1 to 10,000 km² [37] and [57]. There are some indicators that are more flexible and provide reasonable results at both of the considered scales. For example, the sustained yield and the underlying primary productivity can be scaled up or down for largely homogenous ecosystems, such as those in forestry, where sustainable harvest levels or wood resources and residues are common indicators [32].

RESULTS AND DISCUSSION

Major Characteristics of Certification Schemes

The major characteristics evaluated in this study are those identified as relevant by existing reviews [10],[13], [76] and [77], and the evaluated certification schemes and their indicators are introduced in the following sections.

Table 3 shows that only GBEP, NTA 8080, GGL and CSBP target all types of bioenergy. CSBP intends to certify any type of bioenergy from lingo-cellulosic biomass. ISCC, RED cert and RSB originally were developed to demonstrate compliance with national or supra-national legislation, i.e., the EU RED, which primarily cover biofuels and bio liquids [13]. Currently, these schemes are being partially extended and revised to certify solid and gaseous bioenergy to ensure compliance with regulations in potential new versions of the EU RED. NTA 8080 is also used to demonstrate EU RED compliance for biofuels and bio liquids but is the implementation of the "Testing framework for sustainable biomass," the so-called Cramer Criteria, which originally focused on any type and use of sustainable biofuels and other products from biomass [12]. The remaining certification schemes have been developed to ensure sustainable production of agricultural or timber products. To ensure cost-effectiveness, the EU might consider forest certification schemes to be a proof of sustainable production of solid biomass [31]. Table 3 shows that certification schemes for bioenergy attempt to assess the entire supply chain of a product to demonstrate, for example, the higher environmental sustainability than that of fossil energy carriers. The agricultural or forestry certification schemes are rather purpose specific; for example, the schemes demonstrate low-impact cultivation techniques or sustainable forest management [12] and thus focus on feedstock production rather than on the final product. In the latter aspect they differ from bioenergy certification schemes.

Table 3: Major characteristics of certification schemes based on BEFSCI [76], CSBP [43], EC [78], FSC [79], GBEP Task Force [14], GGL [44], GlobalGAP [45], ISCC [46], IWPB [47], Netherlands Standardization Institute [48] and [49], REDcert [50], RSB [51], SAN [52], and SFI [80]

	Bioenergy								Agriculture		Forestry	
	GBEP	NTA8080	ISCC	REDcert	GGL[b]	RSB	CSBP	IWPB	SAN	GlobalGAP	FSC	SFI
Major characteristics:												
Applicable biofuel type												
solid	x	x	(x)	(x)	x	(x)	(x)	x			(x)	(x)
liquid	x	x	x	x	x	x	x					
gaseous	x	x	(x)	(x)	x	(x)	(x)					
Spatial scope for application	Global	Global	Global	EU[a]	Global	Global	US	Global	Global	Global	Regional	US/Canada
EU RED recognition		x	x	x		x						
Degree supply chain coverage[c]	FTPD	FTPD	FTPD	FTPD	FTPD	FTPD	FT	FTPD	F	F	F	F

[a]Few third countries (e.g., Belarus, Ukraine).

[b]Agricultural and forestry source criteria.

[c]Supply chain coverage: feedstock (F), transport (T), processing (P), distribution (D).

Indicator Evaluation

Overview

For the requirements for indicators, the mean of the indicators for certification schemes in Fig. 2 shows that most of the certification schemes are rated at the center of the scale at this aggregation level. The mean for the required resources (assessment interval) with an above-average rating and the mean for the Indicator type with a below-average rating for most of the schemes deviate from the general tendency toward a centered rating.

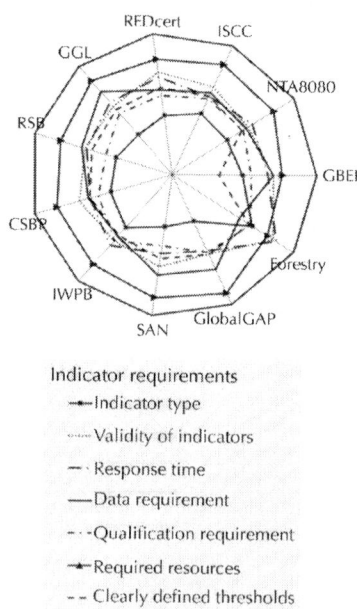

Figure. 2: Arithmetic mean of the ratings by subcategory of the indicator requirements for the evaluated certification schemes and indicator sets, CSBP [43], GBEP Task Force [14], GGL [44], GlobalGAP [45], ISCC [46], IWPB [47], Netherlands Standardization Institute[48] and [49], REDcert [50], RSB [51], SAN [52] and forestry, [29], [53], [54], [55], [56] and [32]; a detailed explanation of the meaning of each step of the rating scales per indicator requirement is discussed in Section 2.2, and the SEM can be found inAppendix A.

The pattern of the Required resources (assessment interval) and Indicator type may be interpreted as the common trade-off between the feasibility and the reliability of indicators (c.f. Payraudeau and van der Werf[37]).

The thematic abundance of indicators not suitable for a direct environmental assessment and therefore excluded for internal consistency of the composite scales has been shown in Section 2.2 in Table 2. Analyzing such excluded indicators gives insight into how certification schemes aim to demonstrate environmental sustainability without an environmental assessment. The majority of the aspects excluded are those not directly related to biomass cultivation or harvesting but are instead related to the handling of equipment and post-production waste or to the documentation of farming activities. The evaluated certification schemes build on cross-compliance or are at least partly set up as a meta-standard. Indicators assess whether legislation or other certification schemes are fulfilled but do not assess whether the environmental impacts of bioenergy production are addressed. Indicators that require the establishment of management plans or actions to achieve a target, such as maintaining water quality, are equally abundant. In minor abundance is the qualification of staff members conducting different tasks in biomass cultivation and processing and generic monitoring activities, such as those related to soil quality.

This overview may provide the impression that the selection of most of the indicators is predominately driven by the aim to allow for highly feasible or practical and probably cost-effective assessment, e.g., leading to assessments that do not require (on-site) measurements, such as demonstrated compliance with local legislation or the review of existing documentation. The named indirect assessment approaches not only consume less time and fewer resources but also do not require an understanding of environmental processes or measurement techniques for an on-site assessment for either the certified party or for the auditor. Certification schemes that require the establishment of generic management plans or monitoring without any consideration of local environmental conditions and processes may facilitate a worldwide sustainability assessment.

Evaluation of Indicators by Requirements and by Composite Scales

The overview in Section 3.2.1 revealed that a high aggregation level does not reveal significant differences between certification schemes. Therefore, the results for the ratings of certification scheme indicators are analyzed at the less aggregated level of composite scales and are grouped by the indicator requirements and their subcategories, see Fig. 3.

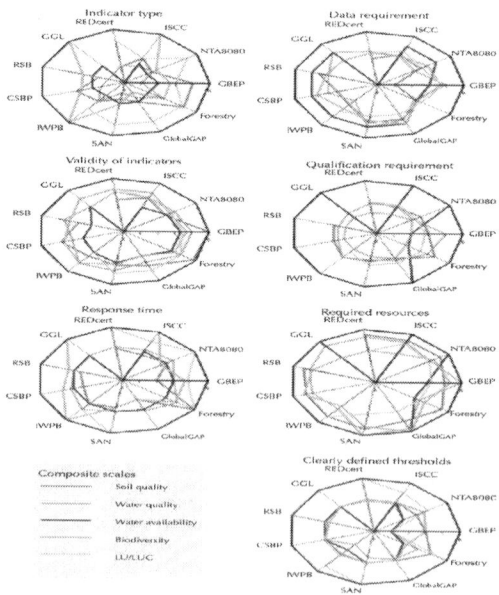

Figure. 3: Arithmetic mean for each indicator requirement subcategory disaggregated by composite scale and certification scheme/indicator set. Five is the best rating; zero indicates a lack of direct environmental assessment indicators for the composite scale and certification scheme. The SEM can be found in Appendix A.

Based on reliability and conceptual soundness, the Indicator type has a nearly universal low rating (value 1–2); i.e., driver indicators on management practices are used, especially for water quality and water availability. Biodiversity and LU/LUC indicators are partially state or impact indicators (value 4–5). These indicators determine whether land use types are converted for biomass production for bioenergy. An example of such state indicators are spatial biodiversity indicators; e.g., there is no bioenergy feedstock production in areas of high conservation value (ecosystems, species). Such indicator demonstrates or intends to demonstrate compliance with EU RED (ISCC, RED cert, IWPB, and GGL). For example, the certification schemes named above assess whether areas of high conservation value or of specific land use types with high carbon stocks, such as peat land, are converted for bioenergy feedstock production. Other EU RED compliance demonstrating schemes (NTA 8080, RSB) without such a pattern have indicators other than spatial indicators that address the protection or restoration of ecological corridors or buffer zones. The Validity of indicators, with the exceptions of the composite scale for water availability and, more significantly, the composite scale other for the non-attributable indicators, could be largely characterized as being validated by models or by agreement in the scientific literature (value 4). The Response time, see Fig. 3, of the chosen indicators is typically one to five years or is not measured, as for causal indicators (value 3), i.e., Indicator type (value 1 or 2). The latter option is more likely because Fig. 3 shows that most of the indicators are causal. Biodiversity and LU/LUC indicators partially show immediate responses (value 5). The rating pattern for Biodiversity and LU/LUC is comparable to the requirements for Indicator type and for the described indicators; see Fig. 3; i.e., the chosen impact indicators are associated with short response times.

Based on the results for feasibility, the Data requirement for the evaluated certification schemes shows that indicators for which data is available at other scales (value 3) or which require data from field observations and questionnaires but measurements (value 4) are not predominately used. The Qualification requirement greatly varies for the different composite scales. The biodiversity indicators are difficult to assess or require prior knowledge. At the least, general higher education, a university degree in agricultural science, or vocational training is required for the assessment (value 2–3). In contrast, the indicators chosen for water availability, e.g., water use per area, require no education or at least no more than a short introduction (value 4–5). The Required resources (assessment interval), soil quality, water quality and availability and other indicators are assessed predominately at intervals longer than one year (value 4) or do not even require field assessment (value 5). Biodiversity and LU/LUC impacts need to be assessed with a higher frequency; some must be assessed annually (value 3). The comparable patterns for Data requirement and required resources (assessment interval) show that the data type and collection mode and the required resources seem to be correlated, i.e., the more effort that data collection for an indicator requires, the higher the frequency of assessment and vice versa. With respect to the requirement clearly defined thresholds, certification schemes mostly only indicate (value 3) how to derive target values/thresholds or use causal indicators. Causal indicators do not require an actual threshold. Instead, the question is whether a (sustainable) management practices is applied or not, i.e., an assessment of compliance or non-compliance. LU/LUC indicators are an exception; for these indicators a threshold is typically given because their formulation implies that there must not be any land conversion for bioenergy feedstock production.

Trade-offs between feasibility (Data requirement, required resources (assessment interval)) and reliability (Indicator type, Response time), mentioned in Section 3.2.1, are especially pronounced for the composite scale for water availability but are also pronounced for soil and water quality. For water availability, the requirements characterizing feasibility, Data requirement and Required resources (assessment interval), are highly rated (value 4 or 5). The Data requirement can be met with field observations or questionnaires (value 4). The Required resources (assessment interval) are minimal because only surveys and no measurements need to be conducted (value 5). Because it is only necessary to complete a survey without measurements and this process requires even less assessment effort than the least frequent measurement, personnel resources and equipment can be saved relative to indicators that are regularly measured.

The indicator requirements for reliability are rated low. Driver indicators (management practices) that measure no response for the Indicator type (value 1–2) and Response time (value 3) are used. Such a trade-off is not pronounced for the Validity of indicators and their feasibility (Data requirement, required resources (assessment interval)) because both are often highly rated (value 4). I.e., many driver indicators are either validated by models or are widely accepted in the scientific literature. The latter explanation applies to many of the indicators in this study. The comparable high ratings for the Data requirement and required resources (assessment interval) reveal that certification schemes preferably use feasible indicators.

In Fig. 4, the results for the rating of certification scheme indicators are grouped by composite scale to reveal possible further patterns.

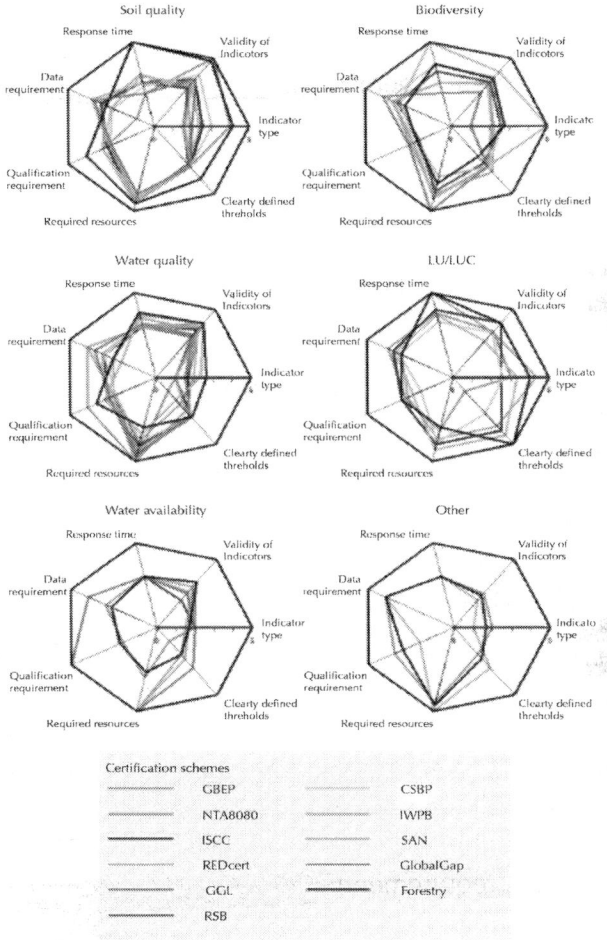

Figure. 4: Arithmetic mean for each composite scale disaggregated by indicator requirement subcategory and certification scheme/indicator set. Five is the best rating; zero indicates a lack of direct environmental assessment indicators for the indicator requirements and certification scheme. The SEM can be found in Appendix A.

Soil quality

With the exception of the Data requirement, soil quality indicators are especially high rated in the forestry indicator set. For the Data requirement, the forestry indicator set still performs as well as most of the other certification schemes. The higher rating of the forestry indicator set might reveal some potential for improvement in existing bioenergy certification schemes.

Water quality

With respect to water quality, most of the certification schemes perform equally well, with the exception of the Data requirement. Here, the low rating of the Indicator type is very apparent and reflects the dominant use of indicators that assess management practices and not the actual changes in the environmental compartment, i.e., water bodies.

Water availability

Water availability could be characterized as highly feasible (Required resources (assessment interval),Qualification requirement, Data requirement) for most of the certification schemes, with the exception of the forestry schemes and GGL, which have low ratings for all of the requirements. This composite scale shows the differences in how well certification schemes chose indicators that optimize the trade-off between requirements, e.g., reliability and feasibility. I.e., a comparable level of reliability and conceptual soundness (Indicator type, Validity of indicators, Response time) may be achieved with a high or low resource use (Required resources (assessment interval), Qualification requirement, Data requirement).

Biodiversity and LU/LUC

Biodiversity is rated very homogenously by ISCC, REDcert, GGL and IWPB and LU/LUC by REDcert, RSB, SAN, GlobalGAP and forestry indicators. Both groups of certification schemes only use one environmental assessment indicator for biodiversity and for LU/LUC respectively; this indicator is no production of bioenergy feedstocks in areas of high conservational value (ecosystems, species) and no conversion of land use types equivalent to those in the EU RED.

The rather high rating observed, especially for biodiversity, can be explained by the nature of the change because the coupling of biodiversity loss to land-use change facilitates the assessment for most of the requirements. Biodiversity gains higher indicator feasibility and reliability and conceptual soundness from land-use change indicators.

Both the biodiversity and LU/LUC indicators also show the extent to which certification schemes exclusively fulfill and go beyond the underlying legislation. Here, the question is how detailed legislation should define environmental impacts that are to be avoided. Assuming that a large abundance of an indicator in the schemes is equal to the relevance, it can be said that the clear indicator definition by EU RED is suitable. This indicator is also used by other certification schemes than those complying with the EU RED. However, this indicator is most likely not sufficient to comprehensively cover the major environmental impacts if only this legal minimum is assessed by certification schemes. Such clearly defined legislation might even hinder the competition among certification schemes to find an optimal solution for comprehensive detection of environmental impacts.

Other

The following composite scales are not completely assessed by the respective scheme. These certification schemes lack direct environmental assessment indicators for some of the composite scales: soil quality (GGL), water availability (REDcert) and LU/LUC (GGL, CSBP) (value 0). Indicators that do not belong to any composite, i.e., indicators grouped under other, are largely missing. Other indicators only occur in the GBEP, IWPB and forestry schemes, as shown in Fig. 4, and contain only three indicators on sustainable harvest levels, which are predominately related to forestry. If indicators for the different composite scales are missing for a certification scheme, they are either neglected by the respective certification scheme or the scheme uses no direct environmental impact assessment indicators, as described in Section 3.2.1.

Comprehensiveness and Quality of Environmental Indicator Sets

The certification schemes and indicator sets for bioenergy production are mapped to the ESS cascade as described in Section 2.3 and as displayed in Appendix A.

Comprehensiveness of Indicators and Causal Links for System Representation

The comprehensiveness of the system representation in these schemes is shown in Table 4.

Table 4: Comprehensiveness of system representation in certification schemes and indicator sets; better ratings mean that more indicators are covered for the different components of the ESS cascade (Fig. 1 in Section 2.3), i.e., the representation of the function of the affected ecosystem and the used ESS. For causal links, the certification schemes are compared with their peers. The certification scheme with the highest number of causal links has the best rating and is used as a benchmark

Certification schemes	Indicatorsa				Causal links
	Ecosystem structures and processes	Ecosystem capacity	Ecosystem services	Human land use activities	
GBEP	−	−	+/−	+/−	−
NTA8080	−	−	+/−	+	+/−
ISCC	−	−	−	+	−
RED cert	−	−	−	+	+/−
GGL S2	+/−	−	+/−	+	+/−
RSB	+/−	+	+/−	+	+
CSBP	−	−	+/−	+/−	+/−
IWPB	−	−	+/−	+	+/−
SAN	+/−	−	+/−	+	+
Global GAP	−	−	−	+	−
Forestry	+/−	+/−	+/−	+	+/−

Coverage of indicators: >66.6%: +, 33.4–66.5%: +/−, <33.3%; −.

Human land use activities can be identified as the most comprehensively covered component of the ESS cascade for most of the schemes reviewed, except for GBEP and ISCC.

This pattern might be explained by the greater feasibility of assessment rather than the relevance of the biophysical processes; see the less comprehensive coverage of ecosystem structures and processes and ESS and the necessity that certification schemes demonstrate sustainability at a local scale instead of the required assessment at a regional scale for other indicators, and see Fig. 1 in Section 2.3. In contrast, the disproportionately small number of indicators to be assessed at a regional scale renders it very likely that certification schemes miss cumulative effects. Cumulative effects are only harmful if a farming practice is applied throughout a region. For example, a crop and the respective fertilizer and pesticide application might only cause significant impacts on water quality if repeatedly applied within a catchment. This problem is addressed by NTA 8080 and IWPB, which both include indicators for off-site impacts, such as the Biological Oxygen Demand. GBEP has a large share of indicators that are beyond the local scale, but this share can very likely be attributed to its difference in purpose. GBEP indicators have been developed for national assessments [14] rather than for certifying single producers.

Ecosystem capacity is considered in most of the certification schemes; however, in RSB ecosystem capacity is not explicitly considered (yellow color) or is not considered (white color), as shown in Appendix A.

An explanation for the lack of thresholds or target values might be the flexibility required to consider the applicability globally and for multiple feed stocks. The indicators need to be equally applicable to different feed stocks that are grown under various environmental conditions and alongside various ecosystems associated with a large variability in ecosystem capacity. Here, clear target values are neither feasible nor practical. However, a methodology for the derivation of the ecosystem capacity can be given. A positive example is the RSB; see Fig. 5. Usually, a threshold is set for the SOC content for several certification schemes. However, the SOC content is only expected to reveal significant changes from changes in management practices, e.g., tillage regime, after a long time lag of at least five to ten years

[81]. Because the reviewed certification schemes do not consider such a time lag in their certificate, such a threshold for SOC will be unlikely to have an impact on the certification decision. Only severe changes of the SOC content over the respective time frame might have an impact.

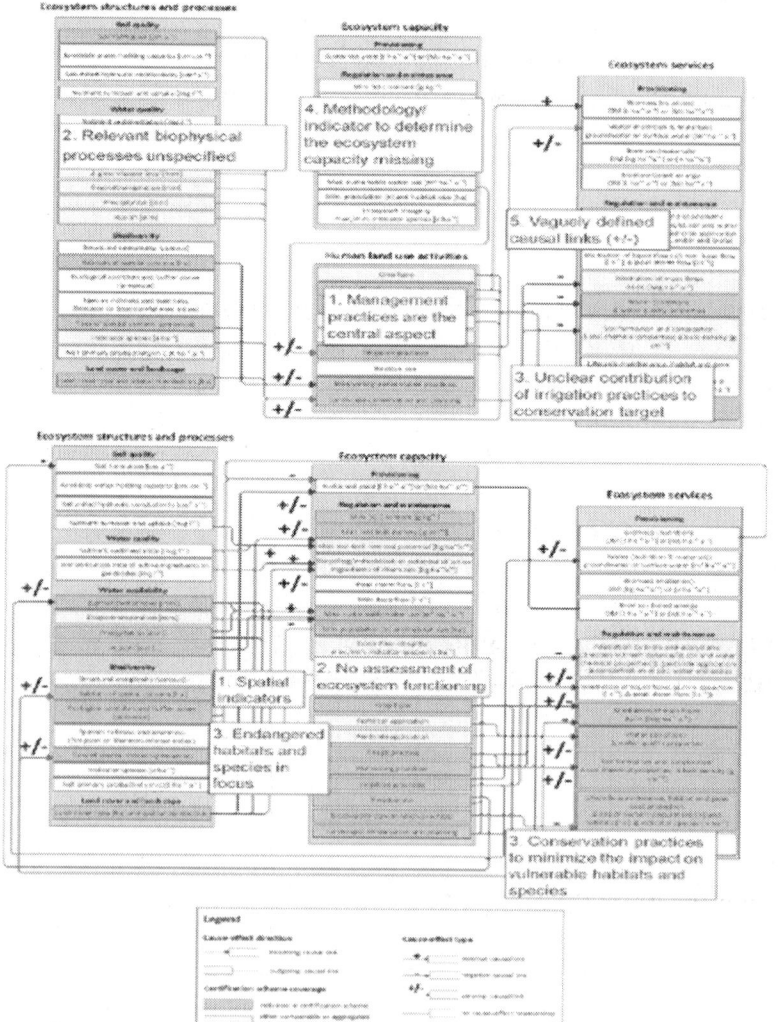

Figure. 5: (upper part) Water availability indicators from GGL mapped onto the ESS cascade. (Lower part) Biodiversity indicators from RSB mapped onto the ESS cascade; common characteristics and deficiencies are indicated in the numbered boxes.

Quality of indicators and causal links for system representation: exemplary cases

The quality of the system representation is analyzed in the examples in Fig. 5; i.e., how certification schemes translate the human–environment interactions and the biophysical cause–effect relationships. As mapped in Fig. 5, the water availability indicators from GGL show that the central aspect of the certification schemes is often driver indicators for management practices, and these indicators should partly consider biophysical processes (2.). These biophysical processes are usually not specified. As an example, indicators are defined as follows: "Data about: climate, water […] are collected on a regular basis." [44]. In addition, it is required that practices are applied to enhance the use of scarce water resources: "4.1 Efficiency and productivity of agricultural water use for better utilization of limited water resources has to increase" [44]. Neither the practices (3.) nor the ecosystem capacity of a scarce water resource (4.) are defined. Missing indicators and open formulations for indicators often result in imprecisely formulated causal links (5.). In contrast to the previous examples, for GBEP, shown in Appendix A, clearly defined indicators, which result in equally clear causal links, can be found.

A higher accuracy of the defined causal links facilitates environmental performance measurements and the determination of options for improvement. Predictions for the alteration of one parameter allow the direction of the change in another indicator to be determined qualitatively or even quantitatively. For example, excluding land cover types such as peat lands from feedstock production reduces the sustainable yield of a region by the theoretical biomass yield of peat land. As shown in Fig. 5, compared with RSB, a deficiency of both GBEP and GGL is the incomprehensive coverage of most of the components of the ESS cascade.

In contrast to GGL, RSB more comprehensively covers the ESS cascade. Despite the greater comprehensiveness, qualitative deficiencies can be shown for examples of the biodiversity indicators from RSB. Preferably, the indicators used are spatial indicators of biodiversity (1.) and not indicators that directly demonstrate ecosystem functioning, such as species richness and evenness indices, e.g., Shannon index, or the abundance of indicator species (2.). The typically chosen spatial indicators and indicators on conservation practices focus on endangered or protected species and habitats (3.).

Possible explanations for the prevailing indicator choice might be:

a. The requirements of the underlying legislations, i.e., the EU RED, govern the indicator choice.

b. Because of their higher risk of extinction, highly vulnerable species and habitats have greater importance for the public or for nature enthusiasts [82].

c. The availability of data for endangered species and habitats is widely available for many parts of the world. Data on species and habitats of less concern is not collected as extensively [68]. Therefore, data availability seems to be better for indicators on endangered species.

f. Indicators on ecosystem function must be adapted to the local context, i.e., indicator species, other indicant of ecosystem functioning and species richness greatly vary by both location and ecosystem.

The most common case in which causal links in certification schemes are defined is when management practices are to be applied to minimize the use of ESS and the creation of disservices is respectively compared with an uncertified alternative in feedstock production. This case is revealed for RSB (3.) and GGL in Fig. 5.

Such an approach neglects the underlying ecosystem structures and processes in the indicator definition. Certification schemes assume a shortened causal link from human land use activities to the ESS and ignore the often directly affected ecosystem structures and processes. Currently, certification schemes are unlikely to allow the measurement and comparison of the environmental performance of bioenergy feed stocks. First, certification schemes, as shown for the example in Fig. 5, partially do not cover the obviously affected ESS. For example, biomass (use) is neglected as an indicator although this indicator could easily be determined. Missing indicators are not only those indicators obtained with more effort or technical skills, such as the impact on the minimum and peak flow of surface waters. Secondly, a large proportion of causal links that are represented by the reviewed certification schemes map the interactions between but not within the different components of the ESS cascade. Therefore, it is not possible to determine trade-offs and synergetic interactions between different ecosystem services.

Thirdly, feedbacks from the use of ESS on ecosystem structures, processes and capacities are mostly not determined, as shown in the mapped certification schemes. Such less comprehensive coverage of the ESS and the causal links renders it impossible to compare the uses and consequently, the environmental impacts of different feed stocks. This deficiency might be because of the nature of the certification schemes to demonstrate compliance with legislation, such as the EU RED or other non-prescriptive rules. The schemes were not originally developed to assess the environmental performances of different feed stocks. Despite this focus, other ESS affected by biomass use could be theoretically used as a multidimensional unit for normalization to allow comparisons of different pathways for biomass provision; this unit would be comparable to the functional unit, e.g., the biomass, in LCAs for energy use or GHG emissions.

Limitations of This Approach

One may argue that there is an assessor bias inherent to both the development and application of the rating scales for the indicator and scheme evaluation. Nevertheless, several measures to reduce and reveal such an assessor bias have been taken:

a.   The use of empirically applied and peer-reviewed rating scales for agri-environmental indicator systems;

b.   The determination of missing rating scales from the range of weak to strong implementation options for bioenergy certification schemes and existing reviews;

c.   Ensuring the transparency of the rating by providing detailed descriptions of each rating scale.

Using the mean to aggregate indicators by composite scale, it was necessary to account for the uncertainty of the mean by the SEM, as shown in Appendix A. There are only a few cases in which the arithmetic mean does not well represent the composite scale. Therefore, the enhanced clarity of the composite scales for each indicator individually should be valued higher. There may be more accurate clustering options than the arithmetic mean, but those options would require complete data sets. Because they do not include indicators for all composite scales, several certification schemes, namely RED cert, GGL, and CSBP, would have had to be excluded. The same problem applies to tests for the internal consistency of the composite scales, such as Cranach's alpha test, which could not be used because the data sets were incomplete. Because only 3 of 87 indicators could not be grouped to the chosen composite scales, as given by the environmental impact categories, the expert-based approach seems to be sufficient.

Empirically, the ESS cascade has been used to assess the impact of human appropriation for purely scientific purposes in a number of cases already, e.g., the studies by Kandziora et al. [64], Maes et al. [63], Petz and van Oudenhoven [61]. Such science-focused studies partially may not reflect practical needs. For example, indicators at the catchment scale are not necessarily suitable to certify individual farmers although these indicators are scientifically more appropriate. In addition, the scope of this study on local/regional environmental impacts required the exclusion of global environmental impacts (e.g., air quality). Therefore, a smaller number of interactions with the related ESS, e.g., the atmospheric composition and climate regulation, are missing. Nevertheless, it is unlikely that a few additional ESS would significantly change the relatively clear patterns shown for the included ESS.

Results in the Context of Existing and Possible Future Research

This section sets the findings of this study in relation to existing research and outlines future research needs.

Usefulness of precise and harmonized legislation on environmental impacts as baseline for certification schemes

Biodiversity and LU/LUC, as composite scales, demonstrate that there is a convergence of certification schemes. The results by Van Dam et al. [13] noting the abundance of spatial biodiversity indicators for endangered habitats and species can be confirmed. The actual change in biodiversity is typically not assessed in the evaluated certification schemes, but it is stated to be hardly possible by current schemes and requiring beyond farm scale assessments [12]. For biodiversity, the hypothesis that precise definitions of the underlying legislation such as the EU RED might hinder the use of more reliable impact indicators seems relevant. In particular, other composite scales with less precise definitions, e.g., the Water Framework Directive in the EU, or with no underlying legislation, such as the scale for water quality, show a larger variety of indicators. Such convergence caused by precisely defined legislation indicates that exclusive peer comparison in existing review papers (e.g., Van Dam et al. [26]) does not completely reveal the limitations and potential improvements.

An additional research-based indicator set, such as the analytical framework developed in this study, revealed further limitations and potential improvements. Based on this analytical framework, limitations in the qualitative and quantitative representations of environmental impacts and the use of ESS in certification schemes could be shown. Some certification schemes are good examples for selected aspects of the assessment of environmental sustainability. Improvements may be achieved by combining the comprehensiveness of RSB with the quality of GBEP, for example. The focus on human land use activity indicators and the largely incomplete assessment of other key functional relationships show that the selection of indicators for certification schemes is driven by feasibility rather than by relevance or reliability. With respect to feasibility, Scarlat and Dallemand [12] recommend striving for a further harmonization of certification schemes through a meta-standard approach or through internationally harmonized minimum sustainability requirements. Their approach might contribute to reduced certification costs, increased feasibility or increased international acceptance of bioenergy certification schemes; these effects are comparable to the developments in forestry certification schemes (e.g., FSC and PEFC). However, enhanced reliability and conceptual soundness of certification schemes require empirical tests or comparisons with a research-based indicator set. The converging biodiversity and LU/LUC indicators have shown some limitations of peer comparison for certification schemes and missing improvement options from academia.

Trade-off between a reliable sustainability assessment and securing feasible compliance with legislation

The focus on feasibility has been apparent in the indicator evaluation in Section 3.2. Existing studies (e.g., Van Dam et al. [13] or Lewandowski and Faaij [2]) identifying the predominant use of feasible causal indicators can be confirmed. Additionally, recent versions of certification schemes, such as the draft from IWPB issued after the findings of former studies, have not been improved in this respect. In addition, the necessity of linking different spatial assessment scales in a proper consideration of environmental impacts has been identified by Van Dam et al. [13]. Nevertheless, this requirement is still only rarely overcome, e.g., by GBEP. With respect to feasibility, Data requirement and required resources could be observed to be drivers for indicator selection. Similarly, the weak inclusion of ecosystem capacities, i.e., thresholds or target values, or the use of causal indicators without thresholds is deficient with respect to both feasibility and conceptual soundness.

Options to improve current certification schemes

The interactions (causal links) between and within the different components of the environmental systems mapped to the ESS cascade often seem to be incomplete and/or only weakly specified; this incompleteness makes quantification of the interactions difficult or even impossible. This limitation could be improved after specification of the causal links. Incomplete indicator sets do not favor the reliable (environmental) performance measurement of feed stocks. Bioenergy certification schemes have been developed to demonstrate compliance rather than to measure and compare the environmental performances of different feed stocks, confirming Diaz-Chavez [29]. In addition, only the compliance or non-compliance with the certification scheme is of interest not the variable degrees of under-/over-compliance of different feed stocks and producers under different environmental conditions. Mostly likely, future certification schemes could consider different degrees of compliance, e.g., different threshold levels, since too high requirements for producers with low financial means may hinder them to participate [2]. Implementation options could be an extension to the current differentiation of mandatory and facultative requirements used in several certification schemes, e.g., NTA 8080. This approach might (i.) raise the information content of certification schemes by visualizing different degrees of environmental performance. (ii.) This approach also facilitates access for small shareholders in developing countries if they initially only need to comply with less strict thresholds. (iii.) This approach could also be used as a strong marketing tool.

CONCLUSIONS

In this study, we evaluated existing indicator sets and certification schemes to assess the environmental sustainability of different feed stocks for bioenergy. No outstanding certification scheme could be identified. Nevertheless, certain available schemes are better than others for assessing the selected environmental impact categories. To date, the proliferation of schemes, which was noted by several authors [12], [13] and [26], has not led to significant changes in the use of reliable and conceptually sound indicators. Instead, schemes strive for feasibility in the indicator choice by complying with existing legislation or consumer expectations. For legislators, potential conclusions could be (i) to require certification schemes and academia to develop more reliable, but still feasible and cost-effective indicator sets, which at least cover the major underlying ecosystem structures and processes, and/or (ii) to consider a methodology to assess the capacity of an ecosystem, i.e., a methodology to determine threshold values for sustainable production. As a second step, certification schemes could assess well-defined causal links and feedbacks for biomass production; for example, schemes could use the adapted versions of the ESS cascade as an analytical framework. The suggested improvements would contribute to increased reliability in the identification of the environmental impacts of bioenergy feedstocks. As an additional benefit, the improved representation of ecosystem functions and feedback mechanisms will facilitate assessments of the interaction between different ESS, such as biomass use, water use or regulating ESS. In further empirical studies, it will be especially interesting to find out, under which conditions cause-related indicators reliably identify sustainable production and for which cases such indicators do not reveal sustainability deficiencies. Beyond the environmental impacts targeted in this study, further social or economic impacts must be considered in bioenergy certification to enable a more comprehensive comparison of alternative feed stocks.

ACKNOWLEDGMENTS

This work was partly funded by the by EU FP7 project SECTOR (282826, http://www.sector-project.eu/) (Production of Solid Sustainable Energy Carriers by Means of Torrefaction). The authors thank Ameur Manceur for advice on statistical methods and five anonymous reviewers for their comprehensive and helpful comments.

REFERENCES

1. Demarrias A. Political, economic and environmental impactsof biofuels: a review. Appl Energy 2009; 86(Suppl. 1):S108e17

2. Lewandowski I, Faaij APC. Steps towards the development of a certification system for sustainable bio-energy trade. Biomass Bioenergy 2006; 30(2):83e104.

3. Ragauskas AJ, Williams CK, Davison BH, Britovsek G, Cairney J, Eckert CA, et al. The path forward for biofuels and biomaterials. Science 2006; 311(5760):484e9.

4. WG III. In: Edenhofer O, Pichs-Madruga R, Sokona Y, Seyboth K, Matschoss P, Kadner S, et al., editors. IPCC special report on renewable energy sources and climate change mitigation. Cambridge and New York: Cambridge University Press; 2012

5. World Commission on Environment and Development. Our common future. Oxford: Oxford University Press; 1987.

6. UN. Johannesburg declaration of sustainable development. World Summit on Sustainable Development. 2002 August 26-September 4. Johannesburg.

7. Cohen B, Winn MI. Market imperfections, opportunity and sustainable entrepreneurship. J Bus Ventura 2007; 22(1):29e49.

8. Hill J, Nelson E, Tillman D, Polasky S, Tiffany D.Environmental, economic, and energetic costs and benefits of biodiesel and ethanol biofuels. Proc Natl Acad Sci U S a 2006; 103(30):11206e10.

9. Robbins M. Policy: fuelling politics. Nature 2011;474(Suppl.7352):S22e4

10. Efroymson RA, Dale VH, Kline KL, McBride AC, Bielicki JM, Smith RL, et al. Environmental indicators of biofuel sustainability: what about context? Environ Manage 2013; McBride AC, Dale VH, Baskaran LM, Downing ME, Eaton LM,

11. Efroymson RA, et al. Indicators to support environmental sustainability of bioenergy systems. Ecol Indic Scarlat N, Dallemand JF. Recent developments of biofuels/

12. Bioenergy sustainability certification: a global overview. Energy Policy 2011; 39(3):1630e46.2011; 11(5): 1277 e8951 (2):291e306.

13. Van Dam J, Junginger M, Faaij APC. From the global efforts o certification of bioenergy towards an integrated approach based on sustainable land use planning. Renew Sustain Energy Rev 2010; 14(9):2445e72.

14. GBEP Task Force. The Global Bioenergy Partnership sustainability indicators for bioenergy. Rome: GBEP Secretariat e FAO; 2011.

15. Von Blottnitz H, Curran MA. A review of assessments conducted on bio-ethanol as a transportation fuel from a net energy, greenhouse gas, and environmental life cycle perspective. J Clean Prod 2007; 15(7):607e19.

16. Haes HAU, Heijungs R, Suh S, Huppes G. Three strategies to overcome the limitations of life-cycle assessment. J Ind Ecol 2004; 8(3):19e32.

17. Koellner T, Geyer R. Global land use impact assessment on biodiversity and ecosystem services in LCA. Int J Life Cycle Assess 2013; 18(6):1185e7.

18. Koellner T, Baan L, Beck T, Branda~o M, Civit B, Margni M, et al. UNEP-SETAC guideline on global land use impact assessment on biodiversity and ecosystem services in LCA. Int J Life Cycle Assess 2013; 18(6):1188e202.

19. Mila` i Canals L, Rigarlsford G, Sim S. Land use impact assessment of margarine. Int J Life Cycle Assess 2013; 18(6):1265e77.

20.   Directive 2013/draft/EU Directive of the European Parliament and of the council on sustainability criteria for solid and gaseous biomass in electricity and/or heating and cooling and biomethane injected in the natural gas network; 2013.

21.   Dale VH, Lowrance R, Mulholland P, Robertson GP. Bioenergy sustainability at the regional scale. Ecol Soc 2010; 15(4):23.

22.   Rametsteiner E, Simula M. Forest certificationdan instrument to promote sustainable forest management? J Environ Manage 2003; 67(1):87e98.

23.   Directive 2009/28/EC. On the promotion of the use of energy from renewable sources and amending and subsequently repealing Directives 2001/77/EC and 2003/30/EC.2009.04.23. Off J Eur Union 2009; L 140:16e62.

24.   Kaphengst T, Ma MS, Schlegel S. At a tipping point? How the debate on biofuel standards sparks innovative ideas for the general future of standardisation and certification schemes. J Clean Prod 2009; 17(Suppl. 1):S99e101.

25.   Junginger M, van Dam J, Zarrilli S, Ali Mohamed F, Marchal D, Faaij A. Opportunities and barriers for international bioenergy trade. Energy Policy 2011; 39(4):2028e42.

26.   Van Dam J, Junginger M, Faaij A, Ju¨ rgens I, Best G, Fritsche U. Overview of recent developments in sustainable biomasscertification. Biomass Bioenergy 2008; 32(8):749e80.

27.   Searchinger T, Heimlich R, Houghton RA, Dong F, Elobeid A, Fabiosa J, et al. Use of US croplands for biofuels increases greenhouse gases through emissions from land-use change. Science 2008; 319(5867):1238e40.

28.   Sikkema R, Steiner M, Junginger M, Hiegl W, Hansen MT, Faaij A. The European wood pellet markets: current status and prospects for 2020. Biofuels Bioprod Bioref 2011; 5(3):250e78.

29.   Diaz-Chavez RA. Assessing biofuels: aiming for sustainable development or complying with the market? Energy Policy 2011; 39(10):5763e9.

30.   Haines-Young R, Potschin M. The links between biodiversity, ecosystem services and human well-being. In: Raffaelli D, Frid C, editors. Ecosystem ecology: a new synthesis. BES Ecological

Reviews Series. C. Cambridge: Cambridge University Press; 2010.

31. Volpi G. EU policy on bioenergy sustainability. [Presentation on the Internet from 5th of December 2012]. Brussels: European Commission; [cited 2013 January 10].

32. Stupak I, Lattimore B, Titus BD, Tattersall Smith C. Criteria and indicators for sustainable forest fuel production and harvesting: a review of current standards for sustainable forest management. Biomass Bioenergy 2011; 35(8):3287e308.

33. Cocchi M, Nikolaisen L, Junginger M, Sheng Goh C, Heinimo¨ J, Bradley D, et al. Global wood pellet industry e market and trade study. Task 40: Sustainable international bioenergy trade. IEA Bioenergy; 2011.

34. Cocchi M, Nikolaisen L, Junginger M, Sheng Goh C, Heinimo¨ J, Bradley D, et al. Global wood pellet industry e market and trade study. Task 40: Sustainable international bioenergy trade. IEA Bioenergy; 2011.

35. Buchholz T, Luzadis VA, Volk TA. Sustainability criteria for bioenergy systems: results from an expert survey. J Clean Prod 2009; 17(Suppl. 1):S86e98.

36. Niemeijer D, de Groot RS. A conceptual framework for selecting environmental indicator sets. Ecol Indic 2008; Payraudeau S, van der Werf HMG. Environmental impact

37. Payraudeau S, van der Werf HMG. Environmental impact assessment for a farming region: a review of methods. Agric Ecosyst Environ 2005; 107(1):1e19.

38. Svarstad H, Petersen LK, Rothman D, Siepel H, Wa¨tzold F. Discursive biases of the environmental research framework DPSIR. Land Use Policy 2008; 25(1):116e25.

39.] Bockstaller C, Gaillard G, Baumgartner D, Freiermuth Knuchel R, Reinsch M, Brauner R, et al. Betriebliches Umweltmanagement in der Landwirtschaft: Vergleich der Methoden INDIGO, KUL/USL, REPRO und SALCA. Colmar: Grenzu¨ berschreitendes Institut zur rentablen umweltgerechten Landbewirtschaftung; 2006.

40. Bockstaller C, Girardin P. How to validate environmental indicators. Agric Syst 2003; 76(2):639e53.

41. Binder CR, Feola G, Steinberger JK. Considering the normative, systemic and procedural dimensions in indicator-based sustainability assessments in agriculture. Environ Impact Assess 2010; 30(2):71e81.

42. Coste J, Guillemin F, Pouchot J, Fermanian J. Methodological approaches to shortening composite measurement scales. J Clin Epidemiol 1997; 50(3):247e52.

43. CSBP. Standard for sustainable production of agricultural biomass. [Monograph on the Internet updated 2012]. Council on Sustainable Biomass Production; [cited 2012 December 21];

44. GGL. Certification requirements Green Gold Label. [Monograph on the Internet updated 2011]. Zwolle: Green Gold Label Foundation; [cited 2012 December 6]; [about 1 screen].

45. GlobalGAP. Integrated farm assurance: all farm base/crops base/combinable crops. [Monographs on the Internet updated 2012]. Cologne: FoodPLUS GmbH; [cited 2012 December 8]; [about 4 screens].

46. ISCC. ISCC EU system documents. [Monographs on the Internet updated 2011]. Cologne: ISCC System GmbH; [cited 2012 December 4]; [about 5 screens].

47. IWPB. IWPB sustainability principles/IWPB sustainability criteria and indicators. [Monographs on the Internet updated 2012]. Linkebeek: Laborelec; [cited 2012 December 21]; [about 3 screens].

48. NTA 8080:2009 NTA 8080 e sustainability criteria for biomass for energy purposes; 2009. pp. 1e41.

49. NTA 8081:2012-04 NTA 8081 e certification scheme for sustainably produced biomass for energy purposes; 2012. pp. 1e27.

50. REDcert. REDcert e EU system documents. [Monographs on the Internet updated 2012]. Bonn: REDcert GmbH [cited 2012 December 12]; [about 3 screens].

51. RSB. RSB sustainability standards e EU RED. [Monographs on the Internet updated 2011]. Chatelaine: Roundtable on Sustainable Biomaterials; [cited 2012 December 8]; [about 5 screens].

52. SAN. Farm Standards/Group Standards. [Monographs on th Internet updated 2011]. San Jose´: Sustainable Agriculture Network; [cited 2012 December 6];.

53. Lattimore B, Smith C, Titus B, Stupak I, Egnell G. Environmental factors in woodfuel production: opportunities, risks, and criteria and indicators for sustainable practices. Biomass Bioenergy 2009; 33(10):1321e42.

54. Newton AC, Kapos V. Biodiversity indicators in national forest inventories. Unasylva 2002; 53(210):56e64.

55.] Schoenholtz SH, Miegroet HV, Burger J. A review of chemical and physical properties as indicators of forest soil quality: challenges and opportunities. For Ecol Manage 2000; 138(1):335e56.

56. Smith GF, Gittings T, Wilson M, French L, Oxbrough A O'donoghue S, et al. Identifying practical indicators of biodiversity for stand-level management of plan.

57. De Groot RS, Alkemade R, Braat L, Hein L, Willemen L. Challenges in integrating the concept of ecosystem services and values in landscape planning, management and decision making. Ecol Compl 2010; 7(3):260e72.

58. MEA Board. In: Hassan R, Scholes R, Ash N, editors. Ecosystems and human well-being: current state and trends findings of the condition and trends working group. Washington, DC: Island Press; 2005.

59. Costanza R, d'Arge R, De Groot R, Farber S, Grasso M, Hannon B, et al. The value of the world's ecosystem services and natural capital. Nature 1997; 387(6630):253e60.

60. Potschin MB, Haines-Young RH. Ecosystem services: exploring a geographical perspective. Prog Phys Geogr 2011; 35(5):575e94.

61. Petz K, van Oudenhoven APE. Modelling land management effect on ecosystem functions and services: a study in the Netherlands. Int J Biodivers Sci Eco Services Manage 2012; 8(1e2):135e55.

62. Van Oudenhoven APE, Petz K, Alkemade R, Hein L, de Groot RS. Framework for systematic indicator selection to assess effects of land management on ecosystem services. Ecol Indic 2012; 21: 110e22.

63. Maes J, Teller A, Erhard M, Liquete C, Braat L, Berry P, et al Mapping and assessment of ecosystems and their services: an

analytical framework for ecosystem assessments under action 5 of the EU biodiversity strategy to 2020. Luxemburg:Publications Office of the European Union; 2013.

64.   Kandziora M, Burkhard B, Mu¨ ller F. Interactions of ecosystem properties, ecosystem integrity and ecosystem service indicatorsda theoretical matrix exercise. Ecol Indic 2013; 28: 54e78.

65.   Ojima D, Moran E, McConnell W, Stafford Smith M, Laumann G, Morais J, et al. In: Young B, editor. Science plan and implementation strategy. Stockholm: IGBP Secretariat; 2005.

66.   Foley JA, DeFries R, Asner GP, Barford C, Bonan G, Carpenter SR, et al. Global consequences of land use. Science 2005; 309(5734):570e4.

67.   CICES. CICES V 4.3. [Spreadsheet on the Internet updated 2013]. University of Nottingham; Nottingham: [cited 2013 January 15.

68.   Dale VH, Beyeler SC. Challenges in the development and use of ecological indicators. Ecol Indic 2001; 1(1):3e10.

69.   Kienast F, Bolliger J, Potschin M, and de Groot RS, Verburg PH, Heller I, et al. assessing landscape functions with broadscaleenvironmental data: insights gained from a prototype development for Europe. Environ Manage 2009; 44(6):1099e120.

70.   McElhinny C, Gibbons P, Brack C, Bauhus J. Forest and woodland stand structural complexity: its definition and measurement. For Ecol Manage 2005; 218(1):1e24.

71.   Wascher DM. Agri-environmental indicators: for sustainable agriculture in Europe. Tilburg: European Centre for Nature Conservation; 2000.

72.   Sposito GE. Scale dependence and scale invariance in hydrology. Cambridge University Press; 1998.

73.   Turner M, O'Neill R, Gardner R, Milne B. Effects of changing spatial scale on the analysis of landscape pattern. Landsc Ecol 1989; 3(3e4):153e62.

74.   Joa~o E. How scale affects environmental impact assessment. Environ Impact Assess 2002; 22(4):289e310.

75.   Joa~o E. A research agenda for data and scale issues in strategic environmental assessment (SEA). Environ Impact Assess 2007; 27(5):479e91.

76. BEFSCI. A compilation of bioenergy sustainability initiatives. [Monographs (each certification scheme) on the Internet updated 2010-11]. Rome: FAO; [cited 2012 December 10];

77. Vissers P, Paz A, Hanekamp E. How to select a biomass certification scheme? In: Lammers E, Op de Coul M, editors. Netherlands Programmes Sustainable Biomass. Utrecht: NL Agency; 2011.

78. EC. Biofuels e sustainability schemes [Internet]. Brussels: European Commission; [cited 2012 December 20]; [about 5

79. FSC. Forest Stewardship Council Certification e ensuring environmental, social and economic benefits [Internet]. Bonn: FSC International Center gemeinnu"tzige Gesellschaft mbH; [cited 2012 December 12.

80. SFI. Requirements for the SFI 2010-2014 program: standards, rules for label use, procedures and guidance. [Monograph on the Internet updated 2010 Jan]. Washington DC: Sustainable Forestry Initiative Inc.; [cited 2012 December 7]; [about 2 screens.

81. West TO, Post WM. Soil organic carbon sequestration rates by tillage and crop rotation. Soil Sci Soc Am J 2002; 66(6):1930e46.

82. Raudsepp-Hearne C, Peterson G, Bennett E. Ecosystem service bundles for analyzing tradeoffs in diverse landscapes. Proc Natl Acad Sci U S A 2010; 107(11):5242e7.

Chapter 7

Biomass Yield and Energy Balance of a Short-Rotation Poplar Coppice With Multiple Clones on Degraded Land During 16 Years

S.Y. Dillen[a], S.N. Djomo[a], N. Al Afas[a,b], S. Vanbeveren[a], and R. Ceulemans[a]

[a]University of Antwerp, Department of Biology, Research Group Plant and Vegetation Ecology, Universiteitsplein 1, B-2610 Wilrijk, Belgium

[b]Al Baath University, Agriculture Faculty, Homs, P.O. Box 77, Syrian Arab Republic

ABSTRACT

Although poplar short-rotation coppice (SRC) systems as an alternative to fossil fuels have been intensively studied, little is known about their biomass potential during several consecutive harvest cycles. For the very first time, this study reports on aboveground biomass yield and energy balance of a 16-year-old poplar SRC with a mixture of 17 pure species and hybrid populous spp. clones. The plantation established on degraded land in Boom, Belgium, was maintained as a low-energy input system, i.e. no irrigation, no fertilizers and no fungicides were applied. The average dry biomass yield during the fourth rotation was 4.3 ± 3.4 ton ha^{-1} year^{-1} across all clones, but the most productive clones yielded up to 10.5 ton ha^{-1} year^{-1}. After 16 years, stool survival ranged from 6 to 91% among clones. Our results demonstrated the sustained biomass potential and reporting capacity after a severe leaf rust attack and after several harvests of the studied populous nigraand populous trichocarpa clones as opposed to hybrids between populous deltoides and P. trichocarpa which hardly survived the fourth rotation. These findings suggest that pure species might perform better than hybrids under suboptimal conditions, e.g. on degraded lands, throughout several harvest cycles and/or after leaf rust infestations. Despite the relatively low yields, the investigated system on degraded land had a positive energy balance producing 7.9 times more energy than it consumed from cradle to plant gate.

INTRODUCTION

Poplars (Populous spp.) grown under a short-rotation coppice (SRC) regime have been extensively studied in function of bioenergy production [1], [2], [3], [4], [5] and [6]. Decades-long research has led to a solid expertise in many countries and practical experience on growing poplar at high densities (i.e. ≥5000 cuttings per hectare) has been translated in best practice guidelines. Yet, the environmental impacts and economic viability of SRC as an alternative energy source to fossil fuels are still under debate [7], [8], [9] and [10]. The environmental impacts and energy balance of dense poplar plantations are evaluated through life cycle assessment (LCA), although a widely accepted and uniform methodological approach is lacking thus far[11]. The economic viability is assessed by means of life cycle cost and by financial models considering the costs and benefits over the entire lifetime of the plantation.

To avoid carbon emissions from land use change and to limit the loss of biodiversity, several authors suggested the use of degraded lands for bioenergy crops over agricultural lands [8], [12] and [13].

About 15%, i.e. 5404 km^2, of Belgium's total area was considered to be degraded land in 2003 [14]. Growing poplar on degraded lands may help in recultivating degraded lands or in preventing further degradation of such lands. Rental or purchase price of degraded lands is cheaper in comparison with agricultural lands. However, in many cases, the extra work needed to bring in amendments or to prepare the site for growing SRC or other energy crops make them more expensive. Also, productivity and yields of SRC on degraded lands may be lower. This raises the question whether the productivity on degraded lands is so low that electricity generation from a poplar SRC system on such lands becomes inefficient, i.e. the system's energy ratio is less than unity. Although energy balances of SRC-based electricity systems have been extensively researched, no studies of energy balances on degraded lands were reported [11]. Further, field experiments covering the complete life span of poplar SRC on agricultural lands are scarce and even inexistent on degraded lands. The life expectancy is believed to be 20–25 years (including 7–8 harvests for 3-year harvest cycles) without significant yield losses, but it can be markedly affected by plant material, by plantation maintenance, by the presence of pathogens and by the planting density in relation to the harvest frequency [15].

Shorter rotation cycles allow higher planting densities and thus, higher biomass yields per unit land area. Coppicing usually stimulates spring re-growth and apparently avoids replanting costs. When rotation lengths are too short for a given species or genotype, re-growth may be hindered by depletion of the carbohydrate reserves primarily stored in the root system [5]. A recent study covering 12 years of poplar SRC in North Italy investigated the effect of 1-, 2- and 3-year harvest cycles on biomass potential of the commonly used Populous deltoides Bartr. Clone Lux [16]. Under the annual harvesting scheme, most poplar stools were soon exhausted and did not survive the seventh year. On the other hand, highest survival rates and maximum productivity were ascertained in plots with a 3-year harvest cycle. For many years, poplars have been in the first place selected for single-stem growth and straight stem form in traditional breeding and selection programmers [17]. As a result, several commercially available poplar clones may not withstand frequent harvesting or short-rotation cycles without a decrease in productivity or in report capacity.

In this study, we document the biomass yield of a 16-year-old poplar SRC with multiple clones on degraded land, more specifically a former waste disposal site moderately polluted with heavy metals [18]. As far as we are aware, this is the longest running SRC plantation with poplar on degraded land. The plantation was maintained as a low-energy input system (no fertilization, no irrigation and no fungicides). We built on earlier work and compared actual yields with those from earlier rotations [6]. To study the dynamics of biomass yield of a poplar SRC over 16 years, we estimated effects of clone and rotation year as well as their mutual interactions on stool survival, number of shoots and biomass yield. We also estimated the energy ratio of the investigated SRC-based electricity system.

MATERIAL AND METHODS

Site and Experimental Design

The SRC plantation was established on a former waste disposal site covered with a mixture of sand, clay and rubble from nearby areas in Boom, Flanders, Belgium (51°05'N, 04°22'E; 5 m above sea level). The site was moderately polluted with heavy metals and nutrient as well as mineral reserves were moderate in comparison with agricultural soils [3]. The 0.55 ha field site was plowed and harrowed before planting in April 1996. Hardwood cuttings (25-cm long) from selected poplar (Populous) clones were planted at a planting density of 10,000 trees per hectare according to a double-row plant design with alternating inter-row distances of 0.75 and 1.5 m and a spacing of 0.9 m between cuttings within rows. The 17 clones were distributed using a randomized block design with three replicate plots per clone. Each plot consisted of 100 trees or 10 rows by 10 columns, but only a core of 6 rows by 6 columns, i.e. 36 assessment trees, was sampled or studied to avoid border effects. A more detailed description of the plantation and of the soil characteristics and conditions has been provided earlier [3]. Monthly mean values of temperature and precipitation over the course of the experiment (1997–2011) are reported in Fig. 1. The meteorological data were obtained from the Royal Meteorological Institute of Belgium.

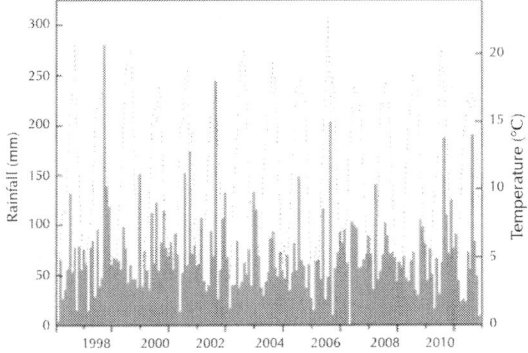

Figure. 1: Average monthly rainfall (mm) and temperature (°C) measured at a meteorological station near the study site, from January 1997 till December 2011.

Plant Material

The planted clones were a mixture of pure species and hybrids: one Populous nigra L. clone (N) Wolterson; three Populus trichocarpa Torr. & Gray clones (T) Columbia River, Fritzi Pauley and Trichobel; six P. trichocarpa × P. deltoides Bartr. Clones (T × D) Beaupré, Boelare, Hazendans, Hoogvorst, Raspalje and Unal; three P. deltoides × P. trichocarpa (D × T) clones IBW1, IBW2 and IBW3; three P. deltoids × P. nigraclones (D × N) Giver, Gibecq and Primo; and one P. trichocarpa × Populus balsamifera L. clone (T × B) Balsam Spire. Place of origin and clonal code numbers have been described by Laureysens et al. [3].

Management Regime

After the establishment year, shoots were manually cut at 5 cm above the ground level in December 1996 to obtain a multi-stem coppice culture. The plantation was harvested in January 2001, in January 2004, in February 2008 and in November 2011. Thus, rotation length was 4 years, except for the second rotation which was only 3 years. Plantation management included mechanical weeding: twice during the establishment year and at the onset of the first rotation, and once at the onset of the three following rotations. Herbicides (glyphosate 3.2 kg ha^{-1} and oxadiazon 9.0 kg ha^{-1}) were applied six times throughout the full life span: twice during the establishment year and once at the onset of each rotation. No irrigation, fertilizers or fungicides were applied. After four rotations, in November 2011, stumps and coarse roots were mechanically removed by an excavator.

Biomass Estimation

Survival of stools (%), number of shoots and shoot diameter were assessed for the 36 assessment trees at the end of the growing season of years 1997–2003, 2005, 2006, 2010 and 2011 [6], [19], [20] and [21]. Shoot diameter (D) was measured at 22 cm above ground level using a digital caliper (Mitutoyo, type CD-15DC, UK). When D exceeded 3 cm, the average of two perpendicular D measurements was further used in the calculations [22]. At regular intervals, a selection of shoots representative of the shoot diameter frequency distribution was randomly harvested from stumps, i.e. 5–30 shoots per clone [19], [20] and [21]. The removal of the shoots was not accounted for in the larger destructive harvests or in the diameter distribution during the next years, since we believe it did not significantly affect the estimations of productivity or total biomass yield. Algometric relationships between shoot dry mass and shoot diameter ($M = {}_aD^b$, with a and b as regression coefficients, and M as shoot dry mass; [22]) were retrieved from a previous study on the same plantation [6]. A previous study at this site demonstrated that one general algometric equation was sufficient for estimating aboveground biomass yield of all clones irrespective of year, except for clone Hazendans and during 2001, a year with severe leaf rust infestation [6]. After each harvest, the total aboveground biomass yield was chipped and transported to the power plant where these chips were gasified for electricity production.

Statistical Analyses

Analyses were performed in the R Statistical Computing Environment (Language Environment Version 2.12.1). Clonal and rotation effects on survival, number of shoots and biomass yield were tested using a repeated measures analysis of variance (ANOVA). The following model was used:

$$z = \mu + cl + y<rt> + rt + cl \times rt + cl \times y<rt> + \square$$

Where z is stool survival, number of shoots or biomass yield; μ is the general mean; clone (cl), rotation (rt) and year (y; nested within rotation) were treated as fixed effects; □ is the residual error. Post-hoc evaluation was done by Tukey›s HSD test. All differences were considered significant at P ≤ 0.05. Pearson correlation coefficients (r) among traits and Spearman rank coefficients ( ) among years were calculated from clonal means.

Energy Analysis

For the studied poplar SRC system, a full chain energy analysis was performed for two situations: (i) from cradle to farm gate and (ii) from cradle to plant gate (Fig. 2). For the latter, the system boundary includes the production of herbicides and tractors, soil cultivation (plowing and harrowing), biomass production, harvest, chipping, storage at the farm, stump removal, transport, natural drying of woody chips and their conversion to electricity (see Fig. 2 for systems boundaries of both situations).

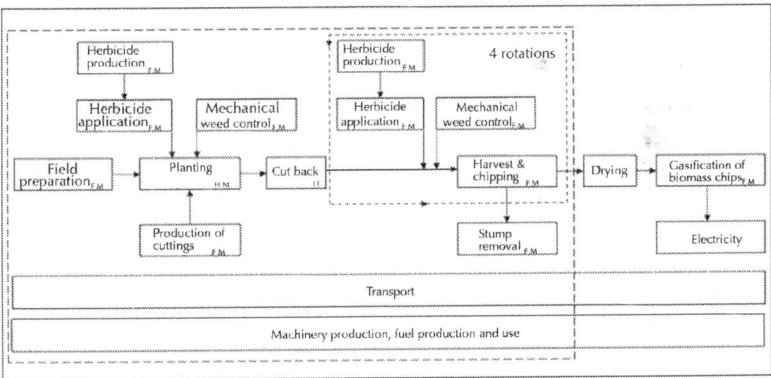

Figure. 2: Schematic representation of the production chain of the studied poplar short-rotation coppice system. All operations are represented by boxes and energy flows by arrows. Inputs of fossil fuel (F), materials (M) and human labor (H) are indicated. Two system boundaries were considered (i) from cradle to farm gate (frame indicated by the dashed lines) and (ii) from cradle to plant gate (frame indicated by the full line). The rotation length was four years, except for the second rotation which was only three years long.

The functional unit was 1 ha land. All direct and indirect energy inputs to the SRC system under study were considered in the inventory up to the production of electricity. Prior to plowing, some works were required to remove rubble from the site, but the energy cost of rubble removal was insignificant and therefore excluded from this analysis. Further, given that the biomass chips were naturally dried, the energy inputs for drying were also excluded from the analysis. Solar energy which initiates the build-up of the poplar trees was not considered in the energy balance. Likewise, an evaluation of environmental impacts was not undertaken.

The direct energy consumption within the system includes the use of diesel or electricity. The indirect energy use involves energy associated with the production of farm machineries and agricultural inputs, such as herbicides and poplar cuttings. Data on material use, diesel consumption, human labor, and machinery used to carry out each agricultural activity were collected onsite (Table 1). To calculate the direct energy costs, we multiplied the amount of diesel consumed during each farming activity by the low heating value of diesel, assumed to be 35.9 MJ l^{-1}[23]. The human energy cost was estimated by multiplying the amount of person-hour of labor for manual planting by the energy expended by a male worker to carry out a farm operation (1.9 MJ h^{-1} [24]). The indirect energy costs of materials were estimated by multiplying the input rate of each material by its energy intensity (Table 1). The assumed energy intensities were 371.1 MJ kg^{-1} for glyphosate [24], 211.2 MJ kg^{-1} for oxadiazon [23], and 0.3 MJ p^{-1} for the cuttings [24]. These values included energy costs for manufacture and transport of the materials to the farm. The indirect energy costs for agricultural machinery production were calculated by multiplying the embodied energy coefficient by the weights of the machine, taking into account the operating rates and life span of the machines (Table 1). For machinery, an embodied energy coefficient of 125 MJ kg^{-1} was assumed [24].

Table 1: Farm activities, material, and fuel inputs for the short-rotation coppice system over 16 years

Activity	Implement used and lifetime (h)	Tractor/Excavator		Total weight (kg)[a]	Operating rate (h ha⁻¹)	Diesel consumed (l ha⁻¹)[b]	Distance (km)	Person-hour of labor (h ha⁻¹)	Number of coverage
		Power (kW)	Lifetime (h)						
Plowing	Moldboard plow (2825)	94	7000	7390	0.86	33.2	–	–	1
Harrowing	Disk tiller (2967)	94	7000	7310	0.82	11.8	–	–	1
Application glyphosate (3.2 kg ha⁻¹)	Boom sprayer (2154)	48	4000	4600	0.37	2.8	–	–	6
Application oxadiazon (9.0 kg ha⁻¹)	Boom sprayer (2154)	48	4000	4600	0.37	2.8	–	–	6
Manual planting (10,000 cuttings ha⁻¹)	–	–	–	–	–	–	–	100	1
Mechanical weeding	Rotortiller (2538)	48	4000	4500	0.44	2.7	–	–	7
Harvest and chipping[c]	Trailer (3000)	94	7000	8200	16.9	74.9	–	–	4
Removal of stumps	Grab bucket crane (5000)	94	9000	22,000	9.5	40.4	–	–	1
Transport of chips to power plant	Truck	–		–	–	–	50	–	4

a The total weight includes weight of implement and weight of tractor.

b The value of diesel consumption refers to an average of all harvests.

c A trailer was used to move the chipping equipment to the field site. A chipper mounted on the trailer was used at the site for the chipping.

To estimate the direct energy costs for the transport of the harvested poplar chips to the conversion site, an energy coefficient of 0.8 MJ ton^{-1} km^{-1} was assumed [24]. We further assumed that the poplar chips were transported by truck over a distance of 50 km, a reasonable distance for a small country like Belgium. The direct energy input for the conversion process itself was estimated at 3% of the energy stored in the woody biomass [25]. The selected conversion technology for this study was a biomass gasification plant with an electrical efficiency of 37.2% [26].

To calculate the total energy input for biomass production we summed up all direct and indirect energy inputs till farm gate. The total energy input at the power plant gate was calculated by adding the energy inputs in conversion plant to the total energy input to produce the biomass. The total energy output at the farm gate was estimated by multiplying the total biomass yield over four rotations by the energy density of wood, i.e. 18.5 MJ kg^{-1} (HHV or higher heating value of poplar wood) [16]. The biomass loss due to natural drying at the farm gate was estimated to be 6% [27], [28], [29] and [30]. We further assumed that no losses occurred during transport and storage of biomass at the gasification plant. At the power plant gate, the total energy output was calculated by multiplying the electrical efficiency by the total biomass energy produced by the SRC system. Finally, we calculated THE cradle to farm gate energy ratio by dividing the harvested biomass energy at the farm gate by the total energy consumed in biomass production (ER_{farm}). In the same way, we calculated the cradle to plant energy ratio by dividing the total energy output at plant gate by the total energy consumed to produce electricity (ER_{plant}).

RESULTS

Biomass Yield

Large clonal variation was observed for stool survival, for number of shoots and for biomass yield (Fig. 3). Striking differences in biomass yields were recorded among the 17 pure and hybrid poplar clones, ranging from 0 to 10.5 ton ha^{-1} year^{-1} during the fourth rotation (2008–2011). While some clones did not survive earlier rotations, other clones displayed highest productivity levels over their entire lifetime (Fig. 3). The pure species clones Wolterson (N), Columbia River (T), Fritzi Pauley (T) and Trichobel (T) were most productive and yielded 8.5–10.5 ton ha^{-1} year^{-1} in the fourth rotation. However, these large yields were attained by fairly contrasting growth strategies. Wolterson produced numerous shoots after coppicing, while the T clones, in particular Fritzi Pauley, accommodated high apical dominance and produced few, but large shoots.

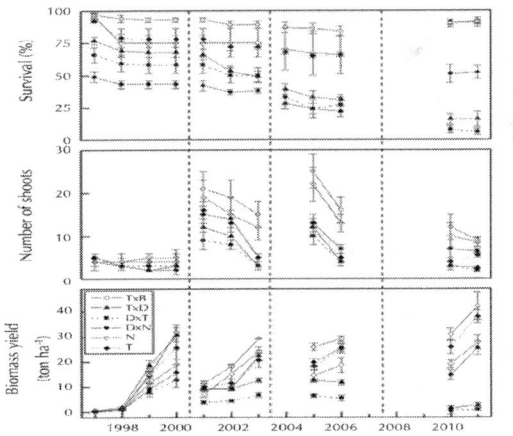

Figure. 3: Time course of survival (%), number of shoots and aboveground dry biomass yield (ton ha^{-1}) during four rotation cycles of the short-rotation coppice culture with 17 poplar clones belonging to six parentages. Means ± standard error are presented; the four rotations are separated by dashed lines. T = Populus trichocarpa; B = P. balsamifera; D = P. deltoides; N = P. nigra.

The performance of some clones varied substantially over different rotations and years. Significant clone × rotation interactions were observed for all studied productivity traits, and for biomass yield there were also significant clone × year<rotation> interactions (Table 2; Fig. 3). The T × D clones were characterized by fast juvenile growth rates and high biomass yield during the first years but due to high-mortality rates the T × D biomass yield dropped drastically from the second rotation onward (Fig. 3). Clone Hoogvorst (T × D) did even not survive the third rotation. As opposed to T × D clones, D × N clones slowly established and had low growth rates during the first growing season (Fig. 3). After the first rotation, biomass yield of the D × N clones steadily increased to intermediately and highly ranked biomass values compared to all other clones in the fourth and third rotations, respectively. Surprisingly, stool survival of some clones was higher in the fourth than in the third rotation (Fig. 3).

Table 2: Tests of fixed effects of the repeated measures three-way ANOVA model for stool survival, number of shoots and biomass yield. Year was treated as a nested factor within rotation. P-values are indicated in bold when non-significant. *** = $P \leq 0.001$

	Clone	Rotation	Year	Clone × year<rotation>	Clone × rotation
Stool survival	$F_{16,383} = 68.5$***	$F_{3,383} = 116.9$***	$F_{8,383} = 4.2$***	$F_{127,383} = 0.17$ **1.00**	$F_{48,383} = 8.5$***
Number of shoots	$F_{16,346} = 26.1$***	$F_{3,346} = 228.0$***	$F_{7,346} = 56.4$***	$F_{110,346} = 1.00$ **0.45**	$F_{47,346} = 4.9$***
Biomass yield	$F_{16,355} = 40.9$***	$F_{3,355} = 29.8$***	$F_{7,355} = 99.4$***	$F_{110,355} = 2.0$***	$F_{47,355} = 9.5$***

Correlations among Traits

Highly significant correlations were found among traits in 2011, i.e. at the end of the fourth rotation. Obviously, high biomass yield was associated with high stool survival ($r = 0.96$ and $P \leq 0.001$). The number of shoots was also strongly and positively correlated with stool survival ($r = 0.86$ and $P \leq 0.001$). Overall, clones producing a higher number of shoots had higher biomass yield ($r = 0.85$ and $P \leq 0.001$). However, some exceptions were observed: few but larger shoots, resulted in large biomass yields for T clones. According to the Spearman rank coefficients across years (Table 3), clonal stability of biomass yield was generally highest within rotations. Across rotations, the first rotation was not representative for the subsequent rotations, i.e. the first rotation did not provide a proper prediction of the yield of subsequent rotations. Changes in clonal biomass rankings also occurred between the second and the fourth rotation, but Spearman rank coefficients suggested high clonal stability in the last two rotations (Table 3).

Table 3: Spearman rank coefficients calculated from clonal means of biomass yield between the fourth and earlier rotations of the 16-year-old poplar short-rotation coppice system. Years without biomass assessments are put in italics. Significance levels are indicated as follows: *** = $P \leq 0.001$; ** = $P \leq 0.01$; * = $P \leq 0.05$; ns = non-significant

2010		**4th rotation**	
		2011	
1st rotation	1997	ns	ns
	1998	ns	ns
	1999	ns	0.52*
	2000	ns	0.56*
2nd rotation	2001	ns	ns
	2002	0.53*	0.87***
	2003	0.68**	0.85***
3rd rotation	2004		
	2005	0.74***	0.89***
	2006	0.88***	0.86***
	2007		
4th rotation	2010		0.97***

Energy Inputs and Outputs

The total energy input to produce the woody chips over 16 years was 49.3 GJ ha^{-1} while the total energy inputs to produce the usable energy, i.e. electricity was 68.8 GJ ha^{-1} (Fig. 4). Field preparation accounted for 4.6% of the total energy costs from cradle to plant gate. Weeding accounted for 30% of the total energy costs, primarily due to the large energy requirements of the herbicide production (Fig. 4). A similar energy cost was related to the operations of harvesting and chipping, 26.8% of the total energy costs. Biomass conversion into electricity was the largest energy cost of the SRC system under study, 23.8% of the total energy input. Relatively little energy was required for production and planting of cuttings, for transport over 50 km and for stump removal, all in the range of 4–5% of the total energy costs. The total biomass feedstock from the studied SRC system was 84.5 ton ha^{-1} of but was reduced to 79.4 after losses due to harvest and to natural drying. The energy yield at the farm gate was 1469.1 GJ ha^{-1}. After conversion of biomass into electricity, total usable energy produced by the studied SRC system was 546.5 GJ ha^{-1}. The ER$_{farm}$ was 29.8 and was reduced to 7.9 when the biomass was converted into electricity, i.e. ER$_{plant}$ (Fig. 4).

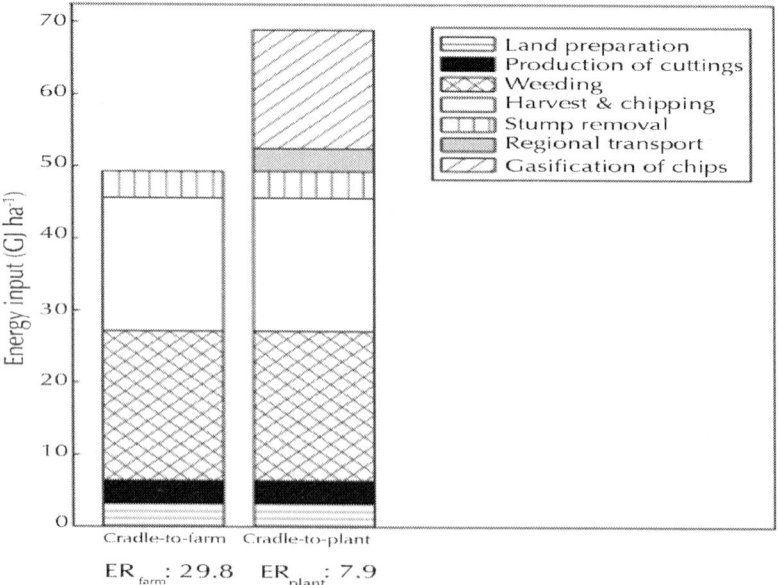

Figure. 4: Breakdown of energy inputs (GJ ha^{-1}) for the poplar short-rotation coppice system during four rotations. Energy costs for each activity and energy ratios (ER) of two system boundaries are presented, i.e. from cradle to farm gate (ER$_{farm}$) and from cradle to plant gate (ERplant). Calculations related to the energy balance are given in Material and Methods.

DISCUSSIONS

Biomass Yield

The average dry biomass yield of 5.3 ton ha^{-1} year^{-1} throughout four rotations is low compared to the frequently reported yields of 10–12 ton ha^{-1} year^{-1}[31]. Nevertheless, significant differences in biomass yields occurred among the planted clones, ranging from 0 to 10.4 ton ha^{-1} year^{-1} during the last rotation. Moreover, the performance of some clones varied substantially over different rotations and years highlighting the need of long-term experiments to identify most suitable poplar clones for SRC. The question whether poplars lose their resprout capacity and growth vigor after several harvests was only partly answered by this study. Clones as Wolterson (N), Fritzi Pauley, Columbia River and Trichobel (T) did not show any sign of stool exhaustion after four harvests and may even have tolerated one or two additional rotations. Indeed, the N and T clones reached peak biomass levels while biomass yields of T × D and D × T clones were lowest after 16 years. For clones of the T, D × N and T × B parentage, higher stool survival was observed in the fourth than in the third rotation. Likely, root sprouts from neighboring trees occupied some of the open areas in the field as new shoots could be distinguished from originally planted individuals indicating the vigorous sprouting capacity of these clones. Breeding and selection for SRC are complex; fast growth rates are not the only aim, but also sustained biomass yields during >20 years and good coppice ability or resprout capacity, i.e. vigorous spring re-growth after coppice [5]. Clones have good coppice ability when their growth is stimulated, or at least, not hampered by frequent harvesting. A large number of shoots after coppice might be considered as an indicator of good coppice ability or resprout capacity as observed for clone Wolterson (N) which produced 20–30 shoots after coppicing and displayed the highest yields during four rotations. Nevertheless, good coppice ability was also observed for studied T clones, all characterized by contrasting growth strategy of few, large shoots.

The poor yields of the D × T and many of the T × D clones can be largely explained by their high susceptibility to leaf rust (Melampsora larici-populina Kleb.) and their intolerance to shade. As previously mentioned [6], a severe rust attack in combination with the bark-killing fungus Discosporium populeum(Sacc. Sutton) during the summer of 2001 reduced the overall yield and caused high mortality, mostly among the D x T and T x D clones. None of these clones completely recovered and their biomass yield continued to decrease, even several years after the major leaf rust infestation. Plots with high mortality as a result of the rust attack were overgrown with tall weeds since weed control was only applied at the onset of each rotation. In the high-mortality plots, the tall weeds likely reduced growth of the resprouting poplars by competing for light, water and/or nutrients. Hybrids usually outperform the pure species at early age and assure rapid establishment of the plantation, particularly hybrids between P. deltoides and P. trichocarpa [32] and [33]. Yet, there seems to be a trade-off between exceptional juvenile growth vigour and tolerance to environmental hazards [6], [34] and [35]. Environmental hazards are most probable to occur throughout the lifetime of a poplar SRC, a period of >20 years. Hence, selection traits as coppice ability as well as tolerance to drought and diseases may be most important in breeding programmers focusing on suitable poplar SRC genotypes. Moreover, this study suggests waiting at least two rotations for poplar breeders to select the most suitable genotypes.

In contrast to monocultures, clonal mixtures tend to reduce yield losses caused by unpredictable environmental changes or hazards [36] and [37]. In the present trial, the clonal mixture appeared to be effective as a disease control strategy; the pure species partly compensated for the losses incurred by the rust infestation. Genetically diverse clones were planted in this (rather small) plantation, i.e. a wide range of pure clones and hybrids of European and North-American species. An intimate mixture of the 17 clones may have been more effective than the actual block design by facilitating a quick occupation of the spaces left by dead stools so that weeds cannot get the upper hand [38]. Although the large heterogeneity of the plantation due to the clonal mixture and block design did not affect harvesting and processing, it did affect plantation maintenance. Particularly weed control was hampered as the poorly yielding or high-mortality plots required more care than the low-mortality plots.

Energy Analysis

The present SRC system yielded an ER_{farm} of 29.8, well within the range of 13–55 presented in a recent review on the energy ratios of poplar SRC [11]. Direct comparison of the present energy budget reported in this study with those from other studies remains complex due to differences in the type of SRC system investigated (low- versus high-input systems), the system boundaries, and the assumptions used. Consistent with previous studies, the use of herbicides as well as harvesting and chipping were the highest energy consumers among the agricultural operations [11]. Our study suggests that poplar SRC grown on degraded lands – in this case moderately polluted with heavy metals – may show very positive energy budgets. Apparently, the relatively low biomass yields throughout the four rotations were compensated for by the low-energy inputs of the system or, in other words, by the absence of irrigation, fertilization and fungicides. Since low inputs imply smaller environmental impacts and lower net carbon dioxide emissions, the studied poplar SRC may be characterized by low environmental impacts and a small contribution to greenhouse gas emissions [39]. However, this and other long-term SRC trials indicated that clonal failures due to diseases and frequent harvesting are likely [6], [16], [40] and [41] advising against the use of constantly high yields in the evaluation of the energy performance of poplar SRC.

Several biomass conversion technologies are readily available, each with their own advantages and disadvantages. In this study, we opted for a biomass gasification plant with an electrical efficiency of 37.2% [26]. Obviously, higher energy efficiencies would be obtained with co-generation of power and heat through this scenario requires a local demand for heat [27]. Like all types of woody biomass, SRC contain heavy metals to some degree, e.g. Pb, Cu and Zn. However, the heavy metal content of SRC from polluted sites may be higher than that of SRC from agricultural lands. Using contaminated enriched SRC for bioenergy purpose can reduce the conversion efficiency [42] or even corrode the boilers [43]. Such risks can be minimized by secondary emission reduction measures, e.g. using filters in boilers [44].

CONCLUSIONS

By growing poplar SRC on degraded lands and with a minimum of energy input (e.g. use of chemicals, irrigation and fertilization), environmental challenges and competition with food crops can be minimized [8]. From this study, we learnt that the SRC systems on degraded lands can payback the energy invested in their production. Carefully selected plant material and adjusted plantation maintenance may even further increase the energy ratio of poplar SRC on degraded lands. Particularly pure P. nigra and P. trichocarpaclones appeared to be most suitable for growth under suboptimal conditions, i.e. being planted on degraded land and coping with several harvest cycles and with diseases as leaf rust. The initially highly promising D × T and T × D hybrids hardly survived the fourth rotation. Therefore, more long-term research is needed to reveal significant shifts in clonal ranking over the entire lifetime of a poplar SRC and to identify most appropriate clones.

ACKNOWLEDGEMENTS

This study was supported by the Flemish Research Foundation (FWO, contract G.0108.97), by the European Commission under the Fourth Framework Programmed (ALTENER, contract AL/95/121/SWE) and under the Seventh Framework programmed (through the European Research Council; ERC Advanced Grant, POPFULL, contract 233366), by the Center of Excellence ECO (University of Antwerp) and by the Province of Antwerp. The project has been carried out in close cooperation with Eta-com B., supplying the premises for the plantation and with the generous support of the city council of Boom. All plant materials were kindly provided by the Research Institute for Nature and Forest (Geraardsbergen, Belgium) and by the Forest Research, Forestry Commission (UK). We are grateful to everyone who helped with biomass yield assessments over the four rotations. S.Y. Dillen is a Research Associate of the Flemish Research Foundation (F.W.O.-Vlaanderen, Belgium).

REFERENCES

1. Ceulemans R, Deraedt W. Production physiology and growth potential of poplars under short-rotation forestry culture. Forest Ecol Manag 1999; 121:9.

2. Kauter D, Lewandowski I, Claupein W. Quantity and quality of harvestable biomass from Populus short-rotation coppice for solid fuel use e a review of the physiological basis and management influences. Biomass Bioenerg 2003; 24:411.

3. Laureysens I, Bogaert J, Blust R, Ceulemans R. Biomass production of 17 poplar clones in a short-rotation coppice culture on a waste disposal site and its relation to soil characteristics. Forest Ecol Manag 2004; 187:295.

4. Dickmann DI. Silviculture and biology of short-rotation woody crops in temperate regions: then and now. Biomass Bioenerg 2006; 30:696.

5. Karp A, Shield I. Bioenergy from plants and the sustainable yield challenge. New Phytol 2008; 179:15.

6. Al Afas N, Marron N, Van Dongen S, Laureysens I, Ceulemans R. Dynamics of biomass production in a poplar coppice culture over three rotations (11 years). Forest Ecol Manag 2008; 255: 1883.

7. Searchinger T, Heimlich R, Houghton RA, Dong FX, Elobeid A, Fabiosa J, et al. Use of US croplands for biofuels increases greenhouse gases through emissions from land-use change. Science 2008; 319:1238.

8. Tilman D, Socolow R, Foley JA, Hill J, Larson E, Lynd L, et al. Beneficial biofuels e the food, energy, and environment trilemma. Science 2009; 325:270.

9. Gasol CM, Brun F, Mosso A, Rieradevall J, Gabarrell X. Economic assessment and comparison of acacia energy crop with annual traditional crops in Southern Europe. Energy Policy 2010; 38:592.

10. Whitaker J, Ludley KE, Rowe R, Taylor G, Howard DC. Sources of variability in greenhouse gas and energy balances for biofuel production: a systematic review. Glob Change Biol Bioenergy 2011; 2:99.

11.   Djomo SN, El Kasmioui O, Ceulemans R. Energy and greenhouse gas balance of bioenergy production from poplar and willow: a review. Glob Change Biol Bioenergy 2011; 3:181.

12.   Fargione J, Hill J, Tilman D, Polasky S, Hawthorne P. Land clearing and the biofuel carbon debt. Science 2008; 319: 1235.

13.   Tilman D, Hill J, Lehman C. Carbon-negative biofuels from low-input high-diversity grassland biomass. Science 2006; 314:1598.

14.   Bai ZG, Dent DL, Olsson L, Schaepman ME. Proxy global assessment of land degradation. Soil Use Manage2008; 24:223.

15.   Sims REH, Maiava TG, Bullock BT. Short-rotation coppice tree species selection for woody biomass production in New Zealand. Biomass Bioenerg 2001; 20:329.

16.   Nassi o Di Nasso N, Guidi W, Ragaglini G, Tozzini C, Bonari E. Biomass production and energy balance of a 12-year-old short-rotation coppice poplar stand under different cutting cycles. Glob Change Biol Bioenergy 2010; 2:89.

17.   Steenackers J, Steenackers M, Steenackers V, Stevens M. Poplar diseases: consequences on growth and wood quality. Biomass Bioenerg 1996; 10:267.

18.   Laureysens I, Blust R, De Temmerman L, Lemmens C, Ceulemans R. Clonal variation in heavy metal accumulation and biomass production in a poplar coppice culture: I. Seasonal variation in leaf, wood and bark concentrations. Environ Pollut 2004; 131:485.

19.   Pellis A, Laureysens I, Ceulemans R. Growth and production of a short-rotation coppice culture of poplar I. Clonal differences in leaf characteristics in relation to biomass production. Biomass Bioenerg 2004; 27:9.

20.   Laureysens I, Pellis A, Willems J, Ceulemans R. Growth and production of a short-rotation coppice culture of poplar. III. Second-rotation results. Biomass Bioenerg 2005; 29:10.

21.   Laureysens I, Deraedt W, Indeherberge T, Ceulemans R. Population dynamics in a 6-year-old coppice culture of poplar. I. Clonal differences in stool mortality, shoot dynamics and shoot diameter distribution in relation to biomass production. Biomass Bioenerg 2003; 24:81.

22. [22] Pontailler JY, Ceulemans R, Guittet J, Mau F. Linear and nonlinear functions of volume index to estimate woody biomass in high-density young poplar stands. Ann Forest Sci 1997; 54:335.

23. Dalgaard T, Halberg N, Porter JR. A model for fossil energy use in Danish agriculture used to compare organic and conventional farming. Agric Ecosyst Environ 2001; 87:51.

24. Frischknecht R, Jungbluth N, Althaus HJ, Bauer C, Doka G, Dones R, et al. Overview and methodology. Ecoinvent report no. 1. Du¨ bendorf, Switzerland: Swiss Centre for Life Cycle Inventory; 2007. p. 68.

25. Garcı´a Cidad V, Mathijs E, Nevens F, Reheul D. Energy crops in Flemish agricultural sector. Report no. 1. Gontrode, Belgium: Support Centre Sustainable Agriculture (Stedula); 2003. p. 94 (in Dutch).

26. Mann MK, Spath PL. Life cycle assessment of a biomass gasification combined-cycle system. Golden, CO, USA: National Renewable Energy Laboratory; December 1997. p. 159.

27. Dubuisson X, Sintzoff I. Energy and CO2 balances in different power generation routes using wood fuel from short-rotation coppice. Biomass Bioenerg 1998; 15:379.

28. Sintzoff I, Martin J, Menu JF, Temmerman M, Thiry J, Tyteca D, et al. Woodsustain. Contributions of wood energy to sustainable development in Belgium. Groupe Energie Biomasse.

29. Nellist ME. Storage and drying of arable coppice. Aspects Appl Biol 1997; 49:349.

30. Borjesson II P. Energy analysis of biomass production and transportation. Biomass Bioenerg 1996; 11:305.

31. Dillen SY, Rood SB, Ceulemans R, In. Growth and physiology. In: Jansson S, Bhalerao R, Groover A, editors. Genetics and genomics of Populus. New York, USA: Springer; 2010. p. 39.

32. Hayes HK. Development of the heterosis concept. In: Gowen JW, editor. Heterosis. Ames, IA, USA: Iowa State College Press; 1952. p. 49.

33. Stettler RF, Zsuffa L, Wu R. The role of hybridization in the genetic manipulation of Populus. In: Stettler RF, Bradshaw HD, Heilman PE, Hinckley TM, editors. Biology of Populus and its implications

for management and conservation. Ottawa, ON, Canada: NRC Research Press, National Research Council of Canada; 1996.

34.   Wu R, Stettler RF. Quantitative genetics of growth and development in Populus. 2. The partitioning of genotype environment interaction in stem growth. Heredity 1997; 78:124.

35.   Dillen SY, Storme V, Marron N, Bastien C, Neyrinck S, Steenackers M, et al. Genomic regions involved in productivity of two interspecific poplar families in Europe. 1. Stem height, circumference and volume. Tree Genet Genomes 2009; 5:147.

36.   DeBell DS, Harrington CA. Deploying genotypes in shortrotation plantations e mixtures and pure cultures of clones and species. Forest Chron 1993; 69:705.

37.   Begley D, McCracken AR, Dawson WM, Watson S. Interaction in short-rotation coppice willow, Salix viminalis genotype mixtures. Biomass Bioenerg 2009; 33:163.

38.   McCracken AR, Dawson WM. Disease effects in mixed varietal plantations of willow. Aspects Appl Biol 2001; 65: 255.

39.   Njakou Djomo S, Blumberga D. Comparative life cycle assessment of three biohydrogen pathways. Bioresour Technol 2011; 102:2684.

40.   Labrecque M, Teodorescu TI. Field performance and biomass production of 12 willow and poplar clones in short-rotation coppice in southern Quebec (Canada). Biomass Bioenerg 2005; 29:1.

41.   Kopp RF, Abrahamson LP, White EH, Volk TA, Nowak CA, Fillhart RC. Willow biomass production during ten successive annual harvests. Biomass Bioenerg 2001; 20:1.

42.   Witters N, Mendelsohn RO, Van Passel S, Van Slycken S, Weyens N, Schreurs E, et al. Phyotremediation, a sustainable remediation technology? II. Economic assessment of CO_2 abatement through the use of phytoremediation crops for renewable energy production. Biomass Bioenerg 2012; 39:470.

43.   Yin C, Rosendahl LA, Kaer SK. Grate-firing of biomass for heat and power production. Progr Energy Combust 2008; 34:725.

44.   Van Loo S, Koppejan J. Handbook of biomass combustion and co-firing. Twente, the Netherlands: Twente University Press; 2002.

Net Ecosystem Production and Carbon Balance of an Src Poplar Plantation During its First Rotation

M.S. Verlinden[a], L.S. Broeckx[a], D. Zona[b], G. Berhongaray[a], T. De Groote[a c], M. Camino Serrano, I.A. Janssen, and R. Ceulemans[a]

[a]Department of Biology, Research Group of Plant and Vegetation Ecology, University of Antwerp, Universiteitsplein 1, B-2610 Wilrijk, Belgium

[b]Department of Animal and Plant Sciences, the University of Sheffield, Western Bank, Sheffield S10 2TN, UK

[c]Unit Environmental Modelling, VITO, Boeretang 200, B-2400 Mol, Belgium

ABSTRACT

To evaluate the potential of woody bioenergy crops as an alternative energy source, there is need for a more comprehensive understanding of their carbon cycling and their allocation patterns throughout the lifespan. We therefore quantified the net ecosystem production (NEP) of a poplar (*Populous*) short rotation coppice (SRC) culture in Flanders during its second growing season.

Eddy covariance (EC) techniques were applied to obtain the annual net ecosystem exchange (NEE) of the plantation. Further, by applying a component-flux-based approach NEP was calculated as the difference between the modelled gross photosynthesis and the respiratory fluxes from foliage, stem and soil obtained via up scaling from chamber measurements. A combination of biomass sampling, inventories and up scaling techniques was used to determine NEP via a pool-change-based approach.

Across the three approaches, the net carbon balance ranged from 96 to 199 g m^{-2} y^{-1} indicating a significant net carbon uptake by the SRC culture. During the establishment year the SRC culture was a net source of carbon to the atmosphere, but already during the second growing season there was a significant net uptake. Both the component-flux-based and pool-change-based approaches resulted in higher values (47–108%) than the EC-estimation of NEE, though the results were comparable considering the considerable and variable uncertainty levels involved in the different approaches. The efficient biomass production – with the highest part of the total carbon uptake allocated to the aboveground wood – led the poplars to counterbalance the soil carbon losses resulting from land use change in a short period of time.

INTRODUCTION

At the present day energy from biomass has gained interest as an alternative for fossil fuels and as a possibility to bring down greenhouse gas emissions [1] and [2]. Land use changes affecting the cycling and storage of carbon (C) in ecosystems [3] are one of the main causes of the increased greenhouse gas levels in the atmosphere [4] and [5]. However, afforestation of abandoned and marginal farmland can enhance ecosystem C storage and potentially counteract the processes of C loss [6]. Within this context the establishment of short rotation coppice (SRC) plantations for bioenergy production has potential for mitigating the rising greenhouse gas levels in the atmosphere [7]. It has been assumed that the net CO_2 emissions from bioenergy cultures are zero [8] or are so-called 'carbon neutral' by taking up as much C during the growth as released upon conversion to energy. However, bioenergy cultures as SRC plantations might act as a CO_2 source, particularly in the short to medium term [9] due to land use changes. During the consequent years they switch to C sinks [8], [10], [11], [12] and [13]. The net C benefit of such plantations is fairly site specific [12]; it is, therefore, important to study and quantify the C cycle of these ecosystems in more detail and to assess their impact on regional C balances [14].

The net C balance of an ecosystem can be assessed by both NEP (net ecosystem production) and NEE (net ecosystem exchange). NEP is the difference between net primary production (NPP) and heterotrophic ecosystem respiration. NPP is the total amount of new organic matter produced during a certain period [15]. It is the difference of the total photosynthetic uptake of CO_2 by the ecosystem – or the gross primary production (GPP) – and the autotrophic ecosystem respiration. NEE is the net CO_2 flux from the ecosystem to the atmosphere [16], which corresponds to the net difference of photosynthetic carbon uptake and the respiration of autotrophs and heterotrophs [17]. NEE equals NEP disregarding sources and sinks for CO_2 not involving conversion to or from organic C [18]. To understand the dynamics of the ecosystem C sinks, it is important to estimate the size of each C pool and to quantify all C fluxes. Studies of the C balance of SRC plantations are rather scarce [19], [14] and [12] and the simultaneous quantification of NEE with eddy covariance techniques, and of NEP with both C flux and C pool assessments were never done before for an SRC culture.

The present study is part of the large-scale POPFULL project [20] which aims to make a full greenhouse gas balance and to investigate the economic and energetic efficiency of an operational SRC culture with poplar. Within this context, the specific objectives of this study were: (i) to quantify the components of the carbon balance of an SRC plantation; (ii) to quantify NEP and determine the sink–source status and (iii) to compare the estimated NEP with NEE measured through eddy covariance techniques. All measurements were performed during the second growth year of the first rotation.

MATERIAL AND METHODS

Site and Plantation Description

The operational POPFULL site is located in Christi, province East-Flanders, Belgium (51°06 44 N, 3°51 02 E). The region is subjected to an oceanic climate with a long-term average annual temperature and precipitation of 9.5 °C and 726 mm, respectively [21]. According to the Belgian soil classification the area forms part of the sandy region with poor natural drainage [22]. The 18.4 ha site was a former agricultural area consisting of croplands (62%; with corn as the most recent cultivated crop in rotation) and extensively grazed pasture (38%). On 7–10 April 2010 an area of 14.5 ha (excluding the headlands) was planted with 12 selected poplar (*Populous*) and three selected willow (*Salix*) genotypes, representing different species and hybrids of *Populus deltoids*, *Populous maximowiczii*, *Populous Ingra*, and *Populous trichocarpa* and *Salix Viminal's*, *Salix dasyclados*, *Salix alba* and *Salix schwerinii*. The present study focuses on the poplar genotypes only. Using a modified leek planting machine 25 cm long dormant and uprooted cuttings were planted in a double-row planting scheme with alternating distances of 0.75 m and 1.50 m between the rows and on average 1.10 m between trees within the rows (plant density of 8000 ha^{-1}). The plantation was designed in large monoclonal blocks of eight double rows wide that covered both types of former land use (cropland and grazed pasture). Each genotype has minimum two and maximum four replicated blocks with row lengths varying from 90 m to 340 m. After two years of growth (2010, GY1 (growth year) and 2011, GY2) the plantation was harvested for the first time on 2–3 February 2012 with commercially available SRC harvesters. Trees continue growing as a coppice culture with multiple shoots per stool in the following two-year-rotations. More details on site conditions, on poplar materials and on the plantation lay-out are found in Brecks et al. [23].

Meteorological Data and Soil Data

A complete set of environmental variables were recorded continuously from June 2010 till present as described in Zone et al. [24] and [25].

Soil temperature was monitored from the surface until 1 m depth; air temperature and relative humidity were collected at about 5.4 m above the surface. All sensors for these measurements were placed in the immediate proximity of an eddy covariance (EC) mast (see below). For more details on the instruments used we refer to Zone et al. [24] and [25].

Quantification of Carbon Pools

Foliage Pool (F)

Leaf litter was collected during leaf fall from early September to December of GY2 in two plots of 5 × 6 trees for each genotype within each former land use type (n = 48). In each plot three perforated litter traps (litter baskets) of 0.57 m × 0.39 m were placed on the ground along a diagonal transect between the rows covering the wide and the narrow inter-row spacing's. Every two weeks the litter traps of each plot were emptied and leaf dry biomass was determined after oven drying at 70 °C. Successively collected leaf biomass was cumulated over time to obtain the yearly produced foliage. C mass fractions of the leaves were determined on a mixed subsample of three randomly selected mature leaves of different individual leaf area and from different tree heights per plot in September of GY2, at the time when maximum leaf area index was reached [23]. Leaves of plots with the same genotype × land use type combination were merged, yielding six leaves per mixed sample. Samples were ground and analyzed by dry combustion with an NC element analyzer (NC-2100 element analyzer, Carlo Reba Instruments, Italy). These C mass fractions were used to quantify the foliage C production per plot. An average foliage C pool value was then calculated by weighing the averages per genotype × land use type combination with their proportional area in the plantation.

Aboveground Woody Biomass Pool (Ste + Br)

Aboveground woody biomass was determined by combining stem diameter inventory data with algometric equations relating woody biomass with stem diameter. Stem diameter at 22 cm above soil level [26] and [27] was measured in December of GY1 and of GY2 for all trees in one row of each monoclonal block. If a tree had multiple stems, every stem was measured within the tree. Missing trees were recorded as zero to correct for the effective tree density. For each genotype an algometric power relationship was established linking aboveground woody (dry) biomass to stem diameter. Based on the diameter distribution at the end of GY2, ten shoots for each genotype were selected for destructive harvest, covering the widest possible diameter range. Shoot diameter (D) at 22 cm was measured with a digital caliper (model CD-15DC, Mitutoyo Corporation, Japan, 0.01 mm precision). The stem was then harvested at 15 cm above soil level, the average harvesting height of the plantation. After determination of dry biomass (DM) of each stem, values were plotted against diameter and fitted as $DM = a \cdot D^b$. for each of the 12 genotypes (with a and b regression coefficients; cfr. Ref. [27]). All 12 power regressions had an R^2 value of more than 97% with a significance p-level < 0.001. For each genotype a mixed subsample of grated wood material of stem (Ste) and branches (Br) of ten trees was used for the analysis of C mass fraction by dry combustion. From diameter inventories and allometric equations, a weighted average of the change in aboveground woody biomass C pool during the second growing season was calculated as the difference between the standing pool after GY2 and GY1.

Belowground Woody Biomass Pool (Cr) and Stump (Stu)

Coarse root woody biomass was determined by excavation of the root system. Because of the high labour intensity, excavation was restricted to genotypes Koster (P. deltoides Marsh × P. nigra L.) and Skado (P. trichocarpa Hook. × P. maximowiczii Henry), selected as most representative for the plantation based on parentage, origin and area coverage in the plantation. For both genotypes five trees of different stem diameters (from 20 mm to 60 mm at 22 cm height) were selected within each of both former land use types. Right after harvest in February 2012, remaining stumps (Stu) and roots were excavated over an area of 1.1 m × 1.125 m (planting distance in the rows × sum of half inter-row distances). All roots within this sampling area were collected, assuming that roots from adjacent trees compensated for roots of the selected tree growing outside the sampled area. Excavation depth was limited to 0.6 m, as very few roots were observed under 0.6 m [28]. Coarse roots

(Ø > 2 mm) were sampled, and total dry biomass of these coarse roots (CR) and the remaining 15 cm high stump was determined after oven drying at 70 °C. Since no significant effect was found for genotype or former land use, all data were pooled; belowground woody biomass and stump biomass were plotted against stem diameter at 22 cm. As for aboveground woody biomass (cfr. Section 2.3.2), an allometric power regression was fitted. From the diameter inventory the average belowground woody biomass and stump biomass pool were estimated for both GY1 and GY2. Since no coarse root turnover was observed, the belowground biomass production of the GY2 was calculated as the standing pool after GY2 minus the standing pool after GY1. Dried root wood was grated for CN-analysis. An average of the C mass fraction was used for calculating the belowground woody C pool.

Fine Root Pool (Fr)

Sequential soil coring was used to determine fine root production in the same two genotypes as for belowground woody biomass determination, i.e. Koster and Skado. From February to November of GY2 the upper 15 cm soil layer was sampled monthly using an 8 cm diameter × 15 cm deep hand-driven corer (Eijkelkamp Agrisearch equipment, The Netherlands) [29]. At every sampling campaign 20 samples were collected for each genotype of which half in the narrow and half in the wide inter-row spacings, randomly spread over the planting area within the former pasture land use type. Fine roots (Ø < 2 mm) were picked from the sample by hand while (i) separating out weed roots from poplar roots, and (ii) sorting poplar roots in dead and living roots. The sorting of dead and living roots was based on the darker colour and the poorer cohesion between the cortex and the periderm of the dead roots [30]. Following washing, fine poplar roots were oven dried at 70 °C for 1–4 days to determine the dry root biomass per soil surface area. Subsamples of dried roots were grinded and analysed for the C mass fraction (NC-2100 element analyzer, Carlo Erba Instruments, Italy). Fine root production during GY2 (F) was estimated using the decision matrix method for sequential coring based on the changes in pools of living and dead roots between successive samplings [31]. There was a significant difference in fine root biomass between wide and narrow inter-row spacings when compared in a t-test [32]. The cumulative (fine) root biomass production over the year was consequently weighted by the area of inter-row spacings, averaged over both genotypes and converted to carbon using the average fine root C mass fraction.

Soil Carbon Pool (S)

Soil C content (S) was assessed before plantation establishment in GY1 (March 2010). Sampling was performed at 110 spatially distributed locations, of which half in each former land use type [23]. Every 15 cm up to a depth of 90 cm an aggregate sample and a bulk density sample were taken by core sampling (Eijkelkamp Agrisearch equipment, The Netherlands). C mass fractions were determined in three replicates per sample by dry combustion (NC-2100 element analyzer, Carlo Erba Instruments, Italy). From C mass fractions and bulk densities, the carbon pool per 15 cm depth interval was calculated and cumulated over 90 cm. The averages per land use type were weighted by their proportion of plantation area to estimate the initial soil C pool. The input of above- and belowground litter from the poplar trees could lead to a soil C enrichment in the long term, as compared to the arable land where crop residues were removed annually during the former land use. Though even in a shorter term an enriching effect in upper soil layers has been observed in a poplar plantation in Italy [33]. However, soil C pool change generally is a very slow process because compared to the total soil C pool the annual changes are relatively small [34]. We therefore assumed that the poplar trees had not significantly changed the soil C pool over the two years.

Quantification of Carbon Fluxes

Soil CO_2 Efflux (R_s)

The CO_2 efflux from the soil (R_s) was monitored by an automated soil CO_2 flux system (LI-8100, LI-COR Biosciences, Lincoln, NE, USA). Sixteen long-term chambers operating as closed systems were connected to an infrared gas analyzer through a multiplexer (LI-8150, LI-COR Biosciences, Lincoln, NE, USA). The 16 chambers were spatially distributed over the plantation covering both former land use types and only two genotypes (Grimminge (*P. deltoides* Marsh. × (*P. trichocarpa* Hook. × *P. deltoides* Marsh.)) and Skado) due to restricted cable lengths. The system was installed at the end of March of GY2 and logged soil CO_2 efflux for each chamber successively every hour until the end of GY2. Soil CO_2 efflux was extrapolated for the period of January–March by a Neural Network analysis based on soil temperature, which was continuously monitored throughout the year (cfr. Section 2.2). Soil CO_2 efflux was independent of the genotype planted, but differed between the two types of former land use [35]. Values of CO_2 efflux were integrated over time and weighted by the proportion of the two former land use types to obtain the plantation average.

Woody Tissue Co$_2$ Efflux (R_{ste+Br})

On both genotypes Grimminge and Skado stem CO_2 efflux was measured during two intensive field campaigns in GY2. The first campaign was carried out during four days from 8 to 12 August of GY2, whereas the second campaign took place from 26 to 30 November of GY2 when trees were in a dormant state. For both genotypes five trees of different diameters (ranging from 29.7 mm to 50.0 mm, and from 29.9 mm to 50.1 mm at 22 cm height for the first and second campaign, respectively) were selected and measured four times during each measuring campaign at different times of the day. The LI-6400XT gas analyzer (LI-COR Biosciences, Lincoln, NE, USA) was used as an open system in combination with a Plexiglas stem cuvette of 17 cm length and with a diameter of 11 cm (cfr. Ref. [36]). The cuvette was assembled on collars, sealed air-tight at approximately 1 m stem height; for each measurement stem diameter at the attachment point was also recorded. The stem cuvette was equipped with an infrared thermocouple to measure stem temperature during each measurement and covered with aluminium foil to avoid possible CO_2-refixation of the bark. Before each measurement, the sample and reference cells of the gas analyzer were matched after the air in the cuvette was allowed to mix and to stabilize for 30 min. five successive measurements were taken and the average was used for further calculations.

Stem CO_2 efflux data were tested for genotype and diameter effects, but no significant effects were found. Consequently data of August – with a temperature range of 17.8–27.4 °C – and data of November – with a temperature range of 8.2–14.1 °C – were pooled to establish two Q_{10} functions, i.e. one for the growing season and one for the dormant period. The following exponential temperature response was fitted (Eq. (1), originating from Ref. [37]): equation (1)

$$SE = E_{10} \cdot Q_{10} \left((T - 10) / 10 \right) \quad \text{SE=E}_{10}$$

(1)

with SE = stem CO_2 efflux, E_{10} = stem CO_2 efflux at a standard temperature of 10 °C, T = temperature and Q_{10} = the change in the rate of stem CO_2 efflux with a 10 °C change in stem temperature. Stem temperature was closely related to air temperature ($p \leq 0.01$), which was logged half-hourly during the year (cfr. Section 2.2). Stem diameter increment was logged (Point Dendrometer ZN11-O_x-WP, Eifel Consulting, Switzerland) during GY2 from March to December and showed a linear increase in diameter from April to September. The average stem surface area was calculated from the weighted average stem diameter and stem height over the plantation (data published in Ref. [23]). The contribution of branches in the aboveground tree structure (data from Ref. [38]) was also included. Combined with the average tree density an estimate of yearly woody tissue CO_2 efflux was obtained.

Foliar Respiration (R_f)

Leaf gas exchange was measured with a portable open-path gas exchange measurement system (LI-6400, LI-COR Biosciences, Lincoln, NE, USA) equipped with a leaf chamber fluorometer (LI-6400-40, LI-COR Biosciences, Lincoln, NE, USA). Four trees of six genotypes (Bakan and Skado (P. trichocarpa Hook. × P. maximowiczii Henry); Grimminge; Koster and Oudenberg (P. deltoides Marsh × P. nigra L.); Wolterson (P. nigra L.)) were selected for the measurements. The first mature leaf of the current-year shoot in the upper canopy was sampled for gas exchange measurements in monthly campaigns from May to September of GY2. Photosynthetic light response curves were obtained by measurements of the net photosynthetic rate at photosynthetic photon flux densities (PPFD) of 1500, 1000, 800, 600, 400, 200, 100 and 0 $\mu mol \ m^{-2} \ s^{-1}$(blue-red LED source type LI-6400-02B, 13% blue light). Leaves were allowed to equilibrate at least 2 min at each step before logging the data. The following conditions were maintained in the cuvette: CO_2 concentration of 400 $\mu mol \ mol^{-1}$, block temperature of 25 °C and vapour pressure deficit of 1.07 kPa ± 0.03. The PPFD response curve, representing the net photosynthetic rate as a function of PPFD, was fitted to the data by a rectangular hyperbola according to Marshall and Biscoe [39] and Thornley and Johnson [40]. Dark respiration at leaf scale (R_{dark}) was obtained from the y-axis intercept (i.e. the net photosynthetic rate at 0 μmol m^{-2} s^{-1}).

The evolution of leaf area was monitored for all 12 genotypes by monthly leaf area index (LAI) measurements from April to November of GY2 (cfr. Ref. [38]). In four replicated measurement plots per former land use type and per genotype LAI was measured indirectly using an LAI-2200 Plant Canopy Analyzer (LI-COR Biosciences, Lincoln, NE, USA). Measurements were taken along two diagonal transects in each plot with the sensor parallel to the rows and perpendicular to the rows, by comparison of above- and below-canopy readings with a 45° view cap. An average LAI over all genotypes was calculated at each measurement time.

No significant seasonal trend in R_{dark} was observed, hence R_{dark} was averaged over time and genotypes. As foliar dark respiration generally decreases from the upper to the lower canopy [41], [42], [43] and [44], this value was multiplied by a factor 0.75 [44]. This factor is the ratio of foliar respiration rates of sunlit leaves in the upper crown to leaves in medium light, representing the main proportion of the canopy for P. deltoides. Since a constant block temperature was set, no R_{dark}-temperature relationship was established. To approach the temperature response of foliar (dark) respiration, a Q_{10} value of 2.1 was used, established for P. deltoidesleaves in a mid-canopy position [44]. The combination of this R_{dark} value with the evolution of LAI during the season resulted in an estimation of the total foliar respiration at ecosystem scale (R_F) for GY2.

Gross Primary Production (GPP)

Gross Primary Production (GPP) was estimated using the terrestrial biosphere model ORCHIDEE (ORganizing Carbon and Hydrology in Dynamic EcosystEms [45]). This process-based model simulating the phenomena of the terrestrial carbon cycle, calculates the C_3 photosynthesis according to Farquhar et al.[46]. The annual GPP was estimated from LAI, from the photosynthetic parameters, i.e. maximum carboxylation rate (V_{cmax}) and maximum electron transport rate (J_{max}), and from a set of meteorological parameters (short and long-wave radiation, precipitation, wind velocity, humidity, temperature and pressure of the air).

Values for the photosynthetic parameters V_{cmax} and J_{max} were obtained through gas exchange measurements. The experimental design was the same and the measuring protocol was similar as the one explained above for the PPFD response curves to determine foliar respiration. The PPFD was fixed at a saturating value of 1500 μmol s^{-1} m^{-2}. Leaves were acclimatized for 10 min at a CO_2 concentration in the leaf cuvette of 400 μmol mol^{-1}, after which the net photosynthetic rate at a sequence of ten different CO_2 concentrations (i.e. 400, 300, 250, 150, 100, 50, 500, 750, 1000 and 1250 μmol mol^{-1}) was measured. Values for V_{cmax} and J_{max} were determined from the A–C_i response curves using the equations of Farquhar et al. [46].

Net Ecosystem Exchange (NEE)

Net Ecosystem Exchange (NEE) was monitored using the eddy covariance (EC) technique. An EC mast was installed in the northeastern part of the plantation in June of GY1 and was continuously operated to the present day. The mast included a sonic anemometer for the measurement of the three-dimensional wind components, wind speed, wind direction, and a closed-path CO_2/H_2O infrared analyzer (LI-7000, LI-COR Biosciences, Lincoln, NE, USA) among others. The CO_2 and sonic wind speed were recorded at 10 Hz using a model CR 5000 data logger (Campbell Scientific, Logan, Utah, USA). Fluxes of CO_2 were calculated using the EdiRe software (R. Clement, University of Edinburgh, UK [47]) and averaged over 30 min. Data were filtered and gap filled after which these data were used to calculate cumulative NEE averaged for GY2. Further details on the EC unit, on the gap filling procedure, and on the flux calculations are described in Zona et al.[25]. Only poplars were included in the footprint of the EC mast.

Carbon Balance

The value of NEE measured through EC was compared with the NEP determined via two different approaches, i.e. the pool-change-based approach and the component-flux-based approach. The sum of the changes in carbon pools of the different plant components represents the bulk of the net primary production (NPP), which is the result of GPP and the autotrophic respiration [16] and [48]. In this study, the total autotrophic respiration (R_{aut}) was calculated as the sum of foliar respiration, stem CO_2 efflux and 40% of the soil CO_2 efflux ($R_{S\ aut}$) representing root respiration [35] and [49]. NEP results from NPP and the heterotrophic respiration (R_{het}), which was taken as the remaining 60% of the total soil CO_2 efflux thereby ignoring respiration from aboveground animals and microbes. NEP was calculated via the pool-change-based approach as:

$$NEP = NPP - Rhet = F + (Ste + Br) + Stu + CR + FR - 0.6.Rs \qquad (2)$$

where all values are expressed in g m^{-2} y^{-1} of carbon. A few minor components of possible C losses were not taken into account for the NEP calculation, i.e. non-CO_2 losses as CO, CH_4, volatile organic compounds (VOCs) to the atmosphere, dissolved organic carbon (DOCs) to deeper soil layers, mycorrhizae, understory weed growth and herbivory. By applying the component-flux-based approach, NEP was calculated as the incoming GPP flux minus the total ecosystem respiration:

$$NEP = GPP - Reco = GPP - (RS + RSte_{+}Br + RF) \qquad (3)$$

Where R_{eco} represents the total ecosystem respiration calculated as the sum of R_S, R_{Ste+Br} and R_F, also expressed in g m^{-2} y^{-1} of carbon.

NEE, which is mostly defined as the measured flux from the ecosystem to the atmosphere has an opposite sign to NEP; a negative NEE means an uptake by the ecosystem. For reasons of consistency with NEP, the positive value of NEE in this study means a net uptake of carbon.

The contribution of NPP within GPP is often termed carbon use efficiency (CUE), being the ratio of the net production to the sum of the net production and the respiratory cost (NPP/[NPP + Respiration]).

RESULTS

As for nearly all terrestrial biomes [50], the largest C pool in the ecosystem was situated in the soil till 90 cm depth (S; Fig. 1a). Among the persisting C pools of the plant system (woody biomass), the highest amount of carbon was stored in the aboveground woody biomass (Fig. 1). The area weighted average over all genotypes of the root:shoot ratio was 0.46 after GY2. The results of the pool-change-based and component-flux-based measurements are graphically presented in Figs. 1b and 2. The NEE for GY2 determined via EC was valued at 96 g m^{-2} y^{-1} carbon uptake, whereas NEP was estimated to be 140 and 199 g m^{-2} y^{-1} through the pool-change-based and component-flux-based approach, respectively. The NEP values estimated through both the component-flux-based and the pool-change-based approaches were 108% and 47% higher, respectively, than the NEE value of the EC. However, considering the magnitude of the components of the NEP calculation (Fig. 2) and the considerable uncertainty levels in the three techniques, results were comparable (asterisks inFig. 2). The positive value of NEP showed that the ecosystem was already a net sink for CO_2 in GY2. Whereas the plantation was still a net C source during the first year GY1 [25], the C assimilation of trees (GPP of 1281 g m^{-2} y^{-1}) in the following year exceeded the absolute value of total R_e. By summing all C pools, NPP was estimated at 493 ± 27 g m^{-2} y^{-1} (average ± standard error) and the total autotrophic respiration (R_{aut}) was estimated at 729 ± 26 g m^{-2} y^{-1}. Furthermore, the sum of R_{aut} and NPP resulted in 1222 ± 37 g m^{-2} y^{-1}, in which they contribute 60% and 40% (CUE), respectively (Table 1). This sum – defined as GPP – was very close (only 4.6% lower) to the simulated GPP from the ORCHIDEE-model using leaf gas exchange measurements.

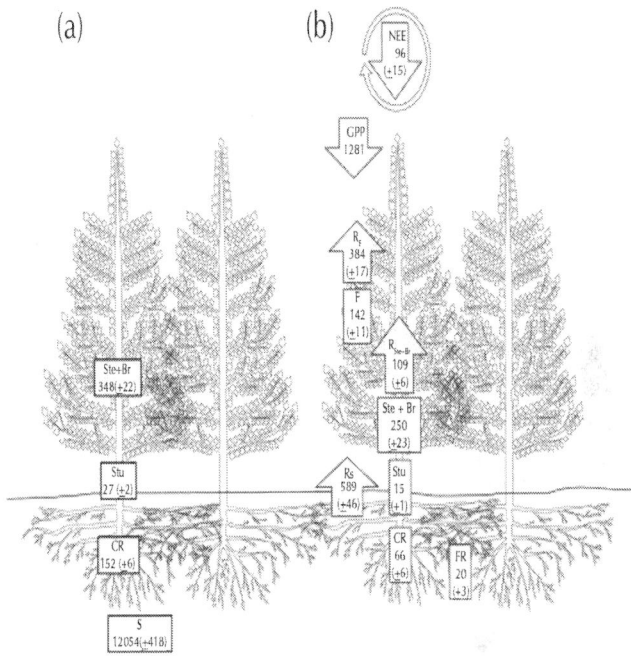

Figure 1: Components and processes of the carbon balance of a high-density poplar plantation during its second year of growth. (a) Bold outlined boxes on the left-hand side represent the standing C pools after two growth years (values in g m^{-2}). (b) Boxes on the right-hand side trees represent annual pool changes, and arrows represent annual integrated C fluxes for the second growing season (values in g m^{-2} y^{-1}). Averaged values are given with standard errors (GPP was a modelled parameter, not including an error range). (Ste + Br) = aboveground woody biomass pool, Stu = aboveground woody stump (15 cm stem) pool remaining after coppicing, CR = coarse root (Ø > 2 mm) pool, S = soil pool till 90 cm depth (determined before plantation establishment), F = foliage pool, FR = fine root (Ø < 2 mm) pool, R_S = total soil CO_2 efflux, R_{Ste+Br} = CO_2 efflux from aboveground woody biomass and R_F = foliar respiration. NEE = net ecosystem exchange measured through the eddy covariance technique (indicated by the circular arrow), which in this case results in a net carbon uptake.

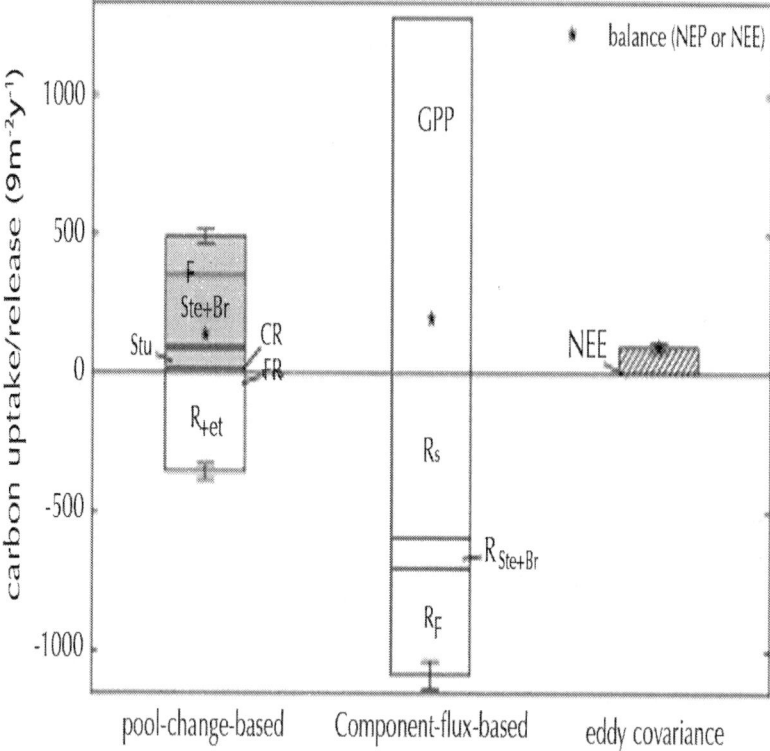

Figure 2: Components of the carbon balance (in g m^{-2} y^{-1}), using three different approaches where uptake and storage are displayed as positive, and release or loss are displayed as negative. The left bar stands for the pool-change-based approach (pool change bars filled in grey); the middle bar stands for the component-flux-based approach (non-filled bars represent integrated fluxes); the right-hand bar represents the eddy covariance measurements (hatched bar). Error bars indicate standard errors (GPP was a modelled parameter, not including an error range). Asterisks show the net result (carbon balance) representing the NEP or NEE for the eddy covariance measurements. For exact values, we refer to the text and to Fig. 1. (Ste + Br) = aboveground woody biomass pool, Stu = aboveground stump (15 cm stem) pool remaining after coppicing, CR = coarse root (Ø > 2 mm) pool, F = foliage pool, FR = fine root (Ø < 2 mm) pool, R_S = total soil CO_2 efflux, R_{het} = heterotrophic soil respiration (60% of R_S), R_{Ste+Br} = CO_2 efflux from aboveground woody biomass and R_F = foliar respiration. NEE = net ecosystem exchange measured through the eddy covariance technique, which in this case results in an uptake.

Table 1: Relative contribution of carbon pool changes to the net primary production (NPP) and the gross primary production (GPP) and relative contribution of fluxes within the total ecosystem respiration (R_{eco}) and GPP. Values are given in percentage of NPP, R_{eco} and GPP. F = foliage pool, (Ste + Br) = aboveground woody biomass pool (stem + branches), Stu = aboveground woody stump (15 cm stem) pool remaining after coppicing, CR = coarse root (Ø > 2 mm) pool, FR = fine root (Ø < 2 mm) pool, R_{Ste+Br} = CO_2 efflux from aboveground woody biomass, R_F = foliar respiration and R_S = the total annual soil CO_2 efflux which is divided in 40% $R_{S\,aut}$ attributed to autotrophic soil (root) respiration and 60% $R_{S\,het}$ as heterotrophic soil respiration

	NPP	R_{eco}	GPP
NPP	100		40.4
F	28.8		11.6
Ste + Br	50.6		20.4
Stu	3.1		1.3
CR	13.3		5.4
FR	4.1		1.7
R_{aut}		67.4	59.6
R_{Ste+Br}		10.1	9.0
R_F		35.5	31.4
$R_{S\,aut}$		21.8	19.3
$R_{het} \approx R_{S\,het}$		32.6	

Soil CO_2 efflux, stem + branch CO_2 efflux and foliar respiration accounted for 54%, 10% and 35% of the total ecosystem respiration, respectively (Table 1). When these three respiratory fluxes were related to GPP, they respectively consumed 46%, 9% and 30% of the total carbon uptake. The remaining 15% of GPP formed NEP. Whereas the aboveground biomass pool showed the highest changes over GY2 it produced the lowest integrated CO_2 efflux. The aboveground biomass pool had the highest share of NPP (51%) followed by the foliage (29%). Both fine (4% of NPP) and coarse roots (13% of NPP) showed the lowest biomass production among all biomass pools – excluding the stump biomass which was considered part of the aboveground woody biomass.

DISCUSSION

Our SRC plantation already represented a significant C sink after two years while other SRC plantations established on agricultural land took two [12] or more than four years [9] before becoming an annual net C sink. During the first years after plantation establishment, crop growth is generally not sufficiently high to compensate for the C losses due to land use change. Several studies showed an initial decrease in the soil C pool during the first years after SRC planting on agricultural soils and grasslands due to intensive mineralization after cultivation [8], [10], [11], [12], [13] and [51]. An integrative study looking at patterns in C cycling across biomes showed that the general trend of negative NEP rates of young (0–10 years) temperate forests was caused by the high heterotrophic respiration rates resulting from disturbance [52]. Likewise, soil C content declined in our plantation during the first two years of growth [35]. However, our present results showed that the growth performance of the poplar canopy counterbalanced this loss already in the second year of growth. Without taking into account other greenhouse gases, this suggests a promising role for SRC with poplar on former agricultural lands with high available nitrogen levels.

Only few studies have assessed the NEP and NPP for a poplar SRC plantation during the first rotation. In a Free Air CO_2 Enrichment (FACE) experiment, Gielen et al. [19] studied net carbon storage in a poplar (*P. nigra* and *Populus euramericana*) SRC ecosystem in central Italy. In the second year after establishment a very high NPP of 1284 g m^{-2} y^{-1} of carbon and an average NEP of 1066 g m^{-2} y^{-1} were reached in the control treatment. Very contrasting results were found for an SRC plantation in Flanders with *P. trichocarpa* × *P. deltoides*, *Salix viminalis*, *Betula pendula* and *Acer pseudoplatanus*, reporting values after the second growth year of 310 g m^{-2} y^{-1} and −360 g m^{-2} y^{-1} for NPP and NEP, respectively [9]. In this last mentioned study no weed control, fertilization or irrigation were applied. Combined with the high heterotrophic soil respiration rate (670 g m^{-2} y^{-1}), this lack of management resulted in a net C source after the second year [9]. In contrast the SRC plantation in the aforementioned FACE experiment was growing in a warmer Mediterranean climate, was irrigated, weeds were removed and herbivores were treated [19]. Moreover, carbon input to the soil was higher than the soil C loss in that study.

Our EC estimate of NEE fell within the range of 70–740 g m^{-2} y^{-1} reported for temperate forests [53]. The differences between the EC result and the NEP estimations, however, are higher than most studies implementing both EC and ecological inventory techniques for estimating NEE and NEP in forest ecosystems. On average, estimates of the net C balance differed between these techniques by 20–30%, although ranging from 7% to 148% [54], [55], [56], [57], [58] and [59]. Moreover, differences between NEP and NEE estimates in the aforementioned studies were not systematic across sites; at some sites NEP seemed to overestimate NEE, while the opposite was true in others.

The quantification of NEE from EC measurements is prone to several uncertainties. The main sources of error of EC measurements are associated with (i) the spatial representativeness of the measured fluxes (footprint issue), (ii) the summation procedure, (iii) the data gap filling and (iv) corrections to night-time data [47]. The precision of the annual integrated EC flux measurements was previously reported as ±5% [60] and [61]; for ideal sites, i.e. extensive canopies on flat terrain, this error bond was set at ±50 g m^{-2} y^{-1} [62]. The uncertainty of the annual NEE flux in our plantation was estimated at 15 g m^{-2} y^{-1} [25], which is, however, much smaller than the differences with NEP estimations. We hypothesize that the component-flux-based approach to determine NEP was the least accurate due to the large uncertainties introduced during upscaling both in space and time. Despite the high precision of foliar CO_2 efflux measurements at leaf scale, crude assumptions were made concerning daily and seasonal evolution which were largely based on growth and temperature, possibly introducing uncertainties in the annual estimates. Spatial uniformity was assumed when scaling up from the leaf to the tree and the stand levels. Similar arguments hold for the woody tissue respiration. A major difficulty involved in measurements of soil, stem and foliar respiration is the respired CO_2 which dissolves in the xylem sap. This portion of respired CO_2 is transported away from the location of production – roots or stem – via the sap flow to upward locations – up to stem, branches and foliage – where it is released to the atmosphere [63], [64], [65] and [66] or possibly fixed by photosynthesis [65]. Consequently, CO_2 efflux measured at a specific location within the tree cannot be considered as the respiration of the measured tissue. Root (and stem) respiration could therefore have been underestimated, whereas stem and foliar respiration could be overestimated. The largest uncertainty involved in the soil CO_2 efflux estimation was the upscaling from a limited number of chambers to the total plantation area (spatial heterogeneity). However, high temporal accuracy was achieved since soil CO_2 efflux was monitored continuously. The relatively simple measures of aboveground C pools (i.e. Ste + Br and F) had a high precision and small aggregation errors since detailed inventories over the whole plantation and among all genotypes were made. Errors in pool change calculations were also limited since the changes are in the same order of magnitude as the pools themselves. Belowground woody biomass had a lower accuracy due to the smaller sample size and the limited number of genotypes that could be sampled. Fine root production is associated with larger uncertainties, which applies in general, regardless the method used [67] and [68]. The largest uncertainty in the pool-change-based approach was, however, the inclusion of the $R_{S\,het}$ which resulted from the partitioning of the total R_S in an autotrophic and a heterotrophic part [35].

A few missing C pools and fluxes, although of minor importance, also hampered closing the carbon balance. Non-CO_2 losses as CO, CH_4 as well as VOCs to the atmosphere, DOCs to deeper soil layers, mycorrhizae, understory weed growth and herbivory were not counted in the NEP calculation. All these carbon balance related processes are usually negligible, but they remain difficult to quantify or to measure [15], [69] and [70]. Small release fluxes of CH_4 were measured at our SRC site [25]. From preliminary results of average DOC concentration and water balance data, the losses of DOCs during GY2 were estimated at 4.7 g m^{-2} y^{-1}, which could be considered as irrelevant with regard to the magnitude of GPP and even NEE. Foliage C losses due to herbivory on the *Populus* trees were estimated at maximal 1% (personal observations), in contrast to the *Salix* species in our plantation, on which we observed substantial infestations of willow beetles (*Phratora vulgatissima*). Emissions of VOCs represented an estimated C loss of 1–2% of GPP of which more than 90% is represented by isoprene (personal communication based on preliminary analysis of PTR–TOF–MS-based flux data; F. Brilli). This C loss corresponded to 13–25 g m^{-2} y^{-1} which was comparable with previous findings for forests [15], [71] and [72]. The understory of herbs dominated by thistles (*Cirsium arvense*) was sparse, and was not quantified in the present study.

At our SRC site fine roots constituted only 4% (Table 1) of the annual NPP whereas for forests it typically ranges from 8% to 76%

[73]. The poor rooting reflected the mesic conditions and the high nitrogen (N) availability of the soil in our plantation, resulting in a lower investment (C allocation) in roots as compared to aboveground biomass [74]. This benefited wood production at an even young plantation age, taking the highest of the total NPP among the different C pools. Foliage had the second largest contribution in NPP. *Populus* trees show an indeterminate growth habit [75] and [76], characterized by continuous shoot growth and leaf production over the growing season. Young developing leaves are net importers of assimilates. When fully expanded they export both acropetally to developing leaves and basipetally to stem and roots until matured; afterwards translocation is mostly to the lower stem and roots [75] and [76]. Mature leaves generally use 20–30% of the C fixed for respiration and maintenance, the remaining 70–80% is exported to developing leaves and stem and roots [77]. Our findings of R_F partitioning for 31% in GPP confirm these general observations. R_F corresponded to half of the autotrophic respiration and was comparable to previous findings in broadleaved forests [78]. This high respiratory cost of foliage could be attributed to the high cellular activity in developing leaves [79]. R_F contributed the second highest within the R_{eco}; R_S took the highest share of R_{eco} of 54% which is slightly lower than the European average of 63% [80]. In forest ecosystems soil carbon efflux is the largest respiratory C flux and following GPP the second most important C flux [81], [82], [83], [84] and [85].

The 40% overall CUE value is within the average range of 39–59% previously reported for temperate deciduous forests [86], [87], [88] and [89]. However, it is much lower than the 60–69% reported for a *Populus* SRC plantation in the second year of growth [19] and lower than the reported average of 58 ± 3% (±st. dev.) for forests with high-nutrient availability [90] which generally invest a larger fraction of GPP to wood compared to forests with low-nutrient availability [86]. In comparison with the low measured R_{Ste+Br} the high share of NPP represented in the aboveground wood production of our SRC plantation (Table 1), suggested an efficient production of wood. This could further justify a high interest in SRC cultures since the wood is the harvestable, and thus economically interesting part.

In conclusion, we were able to quantify all carbon pools and fluxes determining the C balance of this fast-growing SRC culture. The ecosystem was a net carbon sink in the second year of the first rotation, although the results of the different assessment techniques differed in the exact values. The highest respiratory flux was represented by the soil CO_2 efflux whereas the aboveground woody biomass showed the largest carbon pool change.

ACKNOWLEDGEMENTS

This research has received funding from the European Research Council under the European Commission's Seventh Framework Programme (FP7/2007–2013) as ERC grant agreement n°. 233366 (POPFULL), as well as from the Flemish Hercules Foundation as Infrastructure contract ZW09-06. Further funding was provided by the Flemish Methusalem Programme and by the Research Council of the University of Antwerp. G. Berhongaray is a grantee of the Erasmus-Mundus External Cooperation under the EADIC lot 16 Programme; D. Zona was supported by a Marie Curie Reintegration grant (PIRG07-GA-2010-268257) of the European Commission's Seventh Framework Programme; T. De Groote is a PhD fellow of the Research Foundation – Flanders (FWO) and the Flemish Institute for Technological Research (VITO). We gratefully acknowledge the excellent technical support of Joris Cools, the logistic support of Kristof Mouton at the field site, as well as Gerrit Switsers and Nadine Calluy for carbon and nitrogen analyses.

REFERENCES

1.   Buyx, J. Tait Ethical framework for biofuels Science, 332 (2011), pp. 540–541

2.   A.K.M. Sadrul Islam, M. Ahiduzzaman Biomass energy: sustainable solution for greenhouse gas emission AIP Conf Proc, 1440 (2012), pp. 23–32

3.   R.A. Houghton The annual net flux of carbon to the atmosphere from changes in land use 1850–1990 Tellus, 51B (2) (1999), pp. 298–313

4.   E. Kalnay, M. Cai Impact of urbanization and land-use change on climate Nature, 423 (6939) (2003), pp. 528–531

5. R. Lal Soil carbon sequestration to mitigate climate change Geoderma, 123 (1 2) (2004), pp. 1–22 Article PDF (806 K)

6. D.J. Ross, K.R. Tate, N.A. Scott, R.H. Wilde, N.J. Rodda, J.A. Townsend Afforestation of pastures with *Pinus radiata* influences soil carbon and nitrogen pools and mineralisation and microbial properties Aust J Soil Res, 40 (8) (2002), pp. 1303–1318

7. J.I. House, I.C. Prentice, C. Le Quere Maximum impacts of future reforestation or deforestation on atmospheric CO_2 Glob Change Biol, 8 (11) (2002), pp. 1047–1052

8. Soil carbon sequestration beneath hybrid poplar plantations in the north central United States Biomass Bioenergy, 5 (6) (1993), pp. 431–436

9. Vande Walle I. Carbon sequestration in short-rotation forestry plantations and in Belgian forest ecosystems. PhD thesis, Universiteit Gent, Ghent; 2007.

10. D.F. Grigal, W.E. Berguson Soil carbon changes associated with short-rotation systems Biomass Bioenergy, 14 (4) (1998), pp. 371–377

11. H.J. Hellebrand, M. Straehle, V. Scholz, J. Kern Soil carbon, soil nitrate, and soil emissions of nitrous oxide during cultivation of energy crops Nutr Cycl Agroecosyst, 87 (2) (2010), pp. 175–186

12. C.B.M. Arevalo, J.S. Bhatti, S.X. Chang, D. Sidders Land use change effects on ecosystem carbon balance: from agricultural to hybrid poplar plantation Agr Ecosyst Environ, 141 (3–4) (2011), pp. 342–349

13. T. Zenone, J. Chen, M.W. Deal, B. Wilske, P. Jasrotia, J. Xu, *et al*. CO_2 fluxes of transitional bioenergy crops: effect of land conversion during the first year of cultivation Glob Change Biol Bioenergy, 3 (5) (2011), pp. 401–412

14. Vande Walle, R. Samson, B. Looman, K. Verheyen, R. Lemeur Temporal variation and high-resolution spatial heterogeneity in soil CO_2 efflux in a short-rotation tree plantation Tree Physiol, 27 (6) (2007), pp. 837–848

15. D.A. Clark, S. Brown, D.W. Kicklighter, J.Q. Chambers, J.R. Thomlinson, J. Ni Measuring net primary production in forests: concepts and field methods Ecol Appl, 11 (2) (2001), pp. 356–370

16. F.S. Chapin III, G.M. Woodwell, J.T. Randerson, E.B. Rastetter, G.M. Lovett, D.D. Baldocchi, et al. Reconciling carbon-cycle concepts, terminology, and methods Ecosystems, 9 (7) (2006), pp. 1041–1050

17. M. Reichstein, P.C. Stoy, A.R. Desai, G. Lasslop, A.D. Richardson Partitioning of net fluxes M. Aubinet, T. Vesala, D. Papale (Eds.), Eddy covariance: a practical guide to measurement and data analysis, Springer, Dordrecht (2012), pp. 263–289

18. G.M. Lovett, J.J. Cole, M. Pace is net ecosystem production equal to carbon accumulation? Ecosystems, 9 (1) (2006), pp. 1–4

19. B. Gielen, C. Calfapietra, M. Lukac, V.E. Wittig, P. De Angelis, I.A. Janssens, et al. Net carbon storage in a poplar plantation (POPFACE) after three years of free-air CO_2 enrichment Tree Physiol, 25 (11) (2005), pp. 1399–1408

20. POPFULL system analysis of a bio-energy plantation: full greenhouse gas balance and energy accounting [Project Description on the Internet] University of Antwerp, Antwerp (2010) Available from: http://webh01.ua.ac.be/popfull/docs/folder_POPFULL_ENG_comp.pdf [accessed 21.05.13]

21. Klimaat in de wereld klimatogram Gent-Melle. Brussels: Royal Meteorological Institute of Belgium. Available from: http://www.meteo.be/gfx/climatograms/nl6434.png [in Dutch, accessed 21.05.13].

22. E. Van Ranst, C. Sys Eenduidige legende voor de digitale bodemkaart van Vlaanderen (Schaal 1:20 000) Laboratorium voor Bodemkunde, Gent (2000) [in Dutch]

23. L.S. Broeckx, M.S. Verlinden, R. Ceulemans Establishment and two-year growth of a bio-energy plantation with fast-growing Populus trees in Flanders (Belgium): effects of genotype and former land use Biomass Bioenergy, 42 (2012), pp. 151–163

24. Zona D, Janssens IA, Gioli B, Jungkunst HF, Camino Serrano M, Ceulemans R. N_2O fluxes of a bio-energy poplar plantation during a two years rotation period. Glob Change Biol Bioenergy, 2012, http://dx.doi.org/10.1111/gcbb.12019.

25. D. Zona, I.A. Janssens, M. Aubinet, B. Gioli, S. Vicca, R. Fichot, et al. Fluxes of the greenhouse gases (CO_2, CH_4 and N_2O) above a

short-rotation poplar plantation after conversion from agricultural land Agr Forest Meteorol, 169 (2013), pp. 100–110

26. R. Ceulemans, J.Y. Pontailler, F. Mau, J. Guittet Leaf allometry in young poplar stands: reliability of leaf area index estimation, site and clone effects Biomass Bioenergy, 4 (5) (1993), pp. 315–321

27. J.Y. Pontailler, R. Ceulemans, J. Guittet, F. Mau Linear and non-linear functions of volume index to estimate woody biomass in high density young poplar stands Ann For Sci, 54 (4) (1997), pp. 335–34 View Record in Scopus

28. G. Berhongaray, J.S. King, I.A. Janssens, R. Ceulemans An optimized fine root sampling methodology balancing accuracy and time investment Plant Soil, 366 (1–2) (2013), pp. 351–361

29. M.R.G. Oliveira, M. Van Noordwijk, S.R. Gaze, G. Brouwer, S. Bona, G. Mosca, et al. Auger sampling, ingrowth cores and pinboard methods A.L. Smit, A.G. Bengough, C. Engels, M. Van Noordwijk, S. Pellerin, S.C. Van de Geijn (Eds.), Root methods: a handbook, Springer-Verlag, Berlin (2000), pp. 175–210

30. I.A. Janssens, D.A. Sampson, J. Cermak, L. Meiresonne, F. Riguzzi, S. Overloop, et al. Above- and belowground phytomass and carbon storage in a Belgian Scots pine stand Ann For Sci, 56 (2) (1999), pp. 81–90

31. R.I. Fairley, I.J. Alexander Methods of calculating fine root production in forests A.H. Fitter, D. Atkinson, D.J. Read (Eds.), Ecological interactions in soil: plants, microbes and animals, Blackwell Scientific Publications, Oxford (1985), pp. 37–42

32. Berhongaray G, King JS, Janssens IA, Ceulemans R. Fine root biomass and turnover of two fast-growing poplar genotypes in a short-rotation coppice culture. Plant Soil, 2013, http://dx.doi.org/10.1007/s11104-013-1778-x.

33. M.R. Hoosbeek, M. Lukac, D. van Dam, D.L. Godbold, E.J. Velthorst, F.A. Biondi, et al. More new carbon in the mineral soil of a poplar plantation under Free Air Carbon Enrichment (POPFACE): cause of increased priming effect? Global Biochem Cycles, 18 (1) (2004), p. GB1040

34. M. Rodeghiero, A. Heinemeyer, M. Schrumpf, P. Bellamy Determination of soil carbon stocks and changes W.L. Kutsch, M. Bahn, A. Heinemeyer (Eds.), Soil carbon dynamics: an integrated

methodology, Cambridge University Press, Cambridge (2009), pp. 49–75

35. Verlinden MS, Broeckx LS, Wei H, Ceulemans R. Soil CO_2 efflux in a bioenergy plantation with fast-growing *Populus* trees – influence of former land use, inter-row spacing and genotype. Plant Soil, 2013, http://dx.doi.org/10.1007/s11104-013-1604-5.

36. M. Liberloo, P. De Angelis, R. Ceulemans Stem CO_2 efflux of a *Populus nigra* stand: effects of elevated CO_2, fertilization, and shoot size Biol Plant, 52 (2) (2008), pp. 299–306

37. J.H. van't Hoff Lectures on theoretical and physical chemistry. Part I Edward Arnold, London (1898)

38. L.S. Broeckx, M.S. Verlinden, J. Vangronsveld, R. Ceulemans Importance of crown architecture for leaf area index of different *Populus* genotypes in a high-density plantation Tree Physiol, 32 (10) (2012), pp. 1214–122

39. B. Marshall, P. Biscoe A model for C_3 leaves describing the dependence of net photosynthesis on irradiance J Exp Bot, 31 (120) (1980), pp. 29–39

40. J.H.M. Thornley, I.R. Johnson Plant and crop modelling: a mathematical approach to plant and crop physiology Clarendon Press, Oxford (1990)

41. J.R. Brooks, T.M. Hinckley, E.D. Ford, D.G. Sprugel Foliage dark respiration in *Abies amabilis* (Dougl.) Forbes: variation within the canopy Tree Physiol, 9 (3) (1991), pp. 325–338

42. D.S. Ellsworth, P.B. Reich Canopy structure and vertical patterns of photosynthesis and related leaf traits in a deciduous forestOecologia, 96 (2) (1993), pp. 169–178

43. K.A. Mitchell, P.V. Bolstad, J.M. Vose Interspecific and environmentally induced variation in foliar dark respiration among eighteen southeastern deciduous tree species Tree Physiol, 19 (13) (1999), pp. 861–870

44. K.L. Griffin, M. Turnbull, R. Murthy Canopy position affects the temperature response of leaf respiration in *Populus deltoids* New Phytol, 154 (3) (2002), pp. 609–619

45. G. Krinner, N. Viovy, N. de Noblet-Ducoudré, J. Ogée, J. Polcher, P. Friedlingstein, *et al.* A dynamic global vegetation model for

studies of the coupled atmosphere–biosphere system Glob Biogeochem Cycles, 19 (1) (2005), p. GB1015

46. G.D. Farquhar, S. von Caemmerer, J.A. Berry A biochemical model of photosynthetic CO_2 assimilation in leaves of C_3 species Planta, 149 (1) (1980), pp. 78–90

47. M. Aubinet, A. Grelle, A. Ibrom, Ü. Rannik, J. Moncrieff, T. Foken, *et al.* Estimates of the annual net carbon and water exchange of forests: the EUROFLUX methodology Adv Ecol Res, 30 (2000), pp. 113–175

48. S. Luyssaert, M. Reichstein, E.D. Schulze, I.A. Janssens, B.E. Law, D. Papale, *et al.* Toward a consistency cross-check of eddy covariance flux-based and biometric estimates of ecosystem carbon balance Glob Biogeochem Cycles, 23 (3) (2009), p. GB3009

49. P.J. Hanson, N.T. Edwards, C.T. Garten, J.A. Andrews Separating root and soil microbial contributions to soil respiration: a review of methods and observations Biogeochemistry, 48 (1) (2000), pp. 115–146

50. B. Bolin, R. Sukumar, P. Ciais, W. Cramer, P. Jarvis, H. Kheshgi, *et al.* Global perspective R. Watson, I. Noble, B. Bolin, N.H. Ravindranath, D.J. Verardo, D.J. Dokken (Eds.), Land use, land-use change, and forestry – a special report of the IPCC, Cambridge University Press, Cambridge (2000), pp. 23–52

51. P. Nikièma, D.E. Rothstein, R.O. Miller Initial greenhouse gas emissions and nitrogen leaching losses associated with converting pastureland to short-rotation woody bioenergy crops in northern Michigan, USA Biomass Bioenergy, 39 (2012), pp. 413–426

52. K.S. Pregitzer, E.S. Euskirchen Carbon cycling and storage in world forests: biome patterns related to forest age Glob Change Biol, 10 (12) (2004), pp. 2052–2077

53. D. Baldocchi, E. Falge, L. Gu, R. Olson, D. Hollinger, S. Running, *et al.* FLUXNET: a new tool to study the temporal and spatial variability of ecosystem-scale carbon dioxide, water vapor and energy flux densities Bull Am Meteorol Soc, 82 (11) (2001), pp. 2415–2434

54. P.S. Curtis, P.J. Hanson, P. Bolstad, C. Barford, J.C. Randolph, H.P. Schmid, *et al.* Biometric and eddy-covariance based estimates of

annual carbon storage in five eastern North American deciduous forests Agr Forest Meteorol, 113 (1–4) (2002), pp. 3–19

55. J.L. Ehman, H.P. Schmid, C.S.B. Grimmond, J.C. Randolph, P.J. Hanson, C.A. Wayson, et al. An initial intercomparison of micrometeorological and ecological inventory estimates of carbon exchange in a mid-latitude deciduous forest Glob Change Biol, 8 (6) (2002), pp. 575–589

56. K. Black, T. Bolger, P. Davis, M. Nieuwenhuis, B. Reidy, G. Saiz, et al. Inventory and eddy covariance-based estimates of annual carbon sequestration in a Sitka spruce (Picea sitchensis (Bong.) Carr.) forest ecosystem Eur J For Res, 126 (2) (2007), pp. 167–178

57. T. Ohtsuka, W. Mo, T. Satomura, M. Inatomi, H. Koizumi Biometric based carbon flux measurements and net ecosystem production (NEP) in a temperate deciduous broad-leaved forest beneath a flux tower Ecosystems, 10 (2) (2007), pp. 324–334

58. C.M. Gough, C.S. Vogel, H.P. Schmid, H.-B. Su, P.S. Curtis Multi-year convergence of biometric and meteorological estimates of forest carbon storage Agr Forest Meteorol, 148 (2) (2008), pp. 158–170

59. M. Wang, D.-X. Guan, S.-J. Han, J.-L. Wu Comparison of eddy covariance and chamber-based methods for measuring CO_2 flux in a temperate mixed forest Tree Physiol, 30 (1) (2010), pp. 149–163

60. M.L. Goulden, J.W. Munger, S.M. Fan, B.C. Daube, S.C. Wofsy Measurements of carbon sequestration by long-term eddy covariance: methods and a critical evaluation of accuracy Glob Change Biol, 2 (3) (1996), pp. 169–182

61. J.B. Moncrieff, Y. Malhi, R. Leuning The propagation of errors in long-term measurements of land–atmosphere fluxes of carbon and water Glob Change Biol, 2 (3) (1996), pp. 231–240

62. D.D. Baldocchi Assessing the eddy covariance technique for evaluating carbon dioxide exchange rates of ecosystems: past, present and future Glob Change Biol, 9 (4) (2003), pp. 479–492

63. R.O. Teskey, M.A. McGuire Carbon dioxide transport in xylem causes errors in estimation of rates of respiration in stems and branches of trees Plant Cell Environ, 25 (11) (2002), pp. 1571–1577

64.   Saveyn, K. Steppe, M.A. McGuire, R. Lemeur, R.O. Teskey Stem respiration and carbon dioxide efflux of young *Populus deltoides* trees in relation to temperature and xylem carbon dioxide concentration Oecologia, 154 (4) (2008), pp. 637–649

65.   R.O. Teskey, A. Saveyn, K. Steppe, M.A. McGuire Origin, fate and significance of CO_2 in tree stems New Phytol, 177 (1) (2008), pp. 17–32

66.   D.P. Aubrey, R.O. Teskey Root-derived CO_2 efflux via xylem stream rivals soil CO_2 efflux New Phytol, 184 (1) (2009), pp. 35–40

67.   K.A. Vogt, D.J. Vogt, J. Bloomfield Analysis of some direct and indirect methods for estimating root biomass and production of forests at an ecosystem level Plant Soil, 200 (1) (1998), pp. 71–89

68.   I.A. Janssens, D.A. Sampson, J. Curiel-Yuste, A. Carrara, R. Ceulemans The carbon cost of fine root turnover in a Scots pine forest Forest Ecol Manag, 168 (1–3) (2002), pp. 231–240

69.   P. Ciais, M. Wattenbach, N. Vuichard, P. Smith, S.L. Piao, A. Don, *et al.* The European carbon balance. Part 2: Croplands Glob Change Biol, 16 (5) (2010), pp. 1409–1428

70.   P. Smith, G. Lanigan, W. Kutsch, N. Buchmann, W. Eugster, M. Aubinet, *et al.* Measurements necessary for assessing the net ecosystem carbon budget of croplands Agr Ecosyst Environ, 139 (3) (2010), pp. 302–315

71.   J.G. Hamilton, E.H. DeLucia, K. George, S.L. Naidu, A.C. Finzi, W.H. Schlesinger Forest carbon balance under elevated CO_2 Oecologia, 131 (2) (2002), pp. 250–260

72.   Ostonen, K. Lõhmus, K. Pajuste Fine root biomass, production and its proportion of NPP in a fertile middle-aged Norway spruce forest: comparison of soil core and ingrowth core methods For Ecol Manage, 212 (1–3) (2005), pp. 264–277

73.   R.E. Dickson Assimilate distribution and storage A.S. Raghavendra (Ed.), Physiology of trees, John Wiley and Sons Inc., New York (1991), pp. 51–85

74.   A.L. Friend, M.D. Coleman, J.G. Isebrands Carbon allocation to root and shoot systems of woody plants T.D. Davis, B.E. Haissig (Eds.), Biology of adventitious root formation, Plenum Press, New York (1994), pp. 245–273

75. D.R. Geiger Understanding interactions of source and sink regions of plants Plant Physiol Biochem, 25 (5) (1987), pp. 659–666

76. Hagihara, K. Hozumi Respiration A.S. Raghavendra (Ed.), Physiology of trees, John Wiley & Sons Inc., New York (1991), pp. 87–110

77. Tichá, J. Čatský, D. Hodáňová, J. Pospíšilová, M. Kaše, Z. Šesták Gas exchange and dry matter accumulation during leaf development Z. Šesták (Ed.), Photosynthesis during leaf development, Academia, Prague (1985), pp. 157–216

78. I.A. Janssens, H. Lankreijer, G. Matteucci, A.S. Kowalski, N. Buchmann, D. Epron, *et al.* Productivity overshadows temperature in determining soil and ecosystem respiration across European forests Glob Change Biol, 7 (3) (2001), pp. 269–278

79. M.B. Lavigne, M.G. Ryan, D.E. Anderson, D.D. Baldocchi, P.M. Crill, D.R. Fitzjarrald, *et al.* Comparing nocturnal eddy covariance measurements to estimates of ecosystem respiration made by scaling chamber measurements at six coniferous boreal sites J Geophys Res, 102 (D24) (1997), pp. 28977–28985

80. B.E. Law, M.G. Ryan, P.M. Anthoni Seasonal and annual respiration of a ponderosa pine ecosystem Glob Change Biol, 5 (2) (1999), pp. 169–182

81. B. Longdoz, M. Yernaux, M. Aubinet Soil CO_2 efflux measurements in a mixed forest: impact of chamber disturbances, spatial variability and seasonal evolution Glob Change Biol, 6 (8) (2000), pp. 907–917

82. K. Pilegaard, P. Hummelshoj, N.O. Jensen, Z. Chen Two years of continuous CO_2 eddy-flux measurements over a Danish beech forest Agr Forest Meteorol, 107 (1) (2001), pp. 29–41

83. J. Curiel Yuste, M. Nagy, I.A. Janssens, A. Carrara, R. Ceulemans Soil respiration in a mixed temperate forest and its contribution to total ecosystem respiration Tree Physiol, 25 (5) (2005), pp. 609–619

84. R.H. Waring, J.J. Landsberg, M. Williams Net primary production of forests: a constant fraction of gross primary production? Tree Physiol, 18 (2) (1998), pp. 129–134

85. E.H. DeLucia, J.E. Drake, and R.B. Thomas, M. Gonzales-Meler Forest carbon use efficiency: is respiration a constant fraction of

gross primary production? Glob Change Biol, 13 (6) (2007), pp. 1157–1167

86. C.M. Litton, J.W. Raich, M.G. Ryan Carbon allocation in forest ecosystems Glob Change Biol, 13 (10) (2007), pp. 2089–2109

87. M. Campioli, B. Gielen, M. Göckede, D. Papale, O. Bouriaud, A. Granier Temporal variability of the NPP–GPP ratio at seasonal and interannual time scales in a temperate beech forest Biogeosciences, 8 (9) (2011), pp. 2481–2492

88. S. Vicca, S. Luyssaert, J. Penuelas, M. Campioli, F.S. Chapin III, P. Ciais, *et al.* Fertile forests produce biomass more efficiently Ecol Lett, 15 (6) (2012), pp. 520–526.

Chapter 9

Viability Of Off-Grid Electricity Supply Using Ricehusk: A Case Study From South Asia

Subhes C. Bhattacharyya[*]

Institute of Energy and Sustainable Development, De Montfort University, Leicester, UK

ABSTRACT

Rice husk-based electricity generation and supply has been popularized in South Asia by the Husk Power Systems (HPS) and the Decentralised Energy Systems India (DESI), two enterprises that have successfully provided electricity access using this resource. The purpose of this paper is to analyze the conditions under which a small-scale rural power supply business becomes viable and to explore whether larger plants can be used to electrify a cluster of villages. Based on the financial analysis of alternative supply options considering residential and productive demands for electricity under different scenarios, the paper shows that serving low electricity consuming customers alone leads to part capacity utilization of the electricity generation plant and results in a high cost of supply. Higher electricity use improves the financial viability but such consumption behaviour benefits high consuming customers greatly. The integration of rice mill demand, particularly during the off-peak period, with a predominant residential peak demand system improves the viability and brings the levelised cost of supply down. Finally, larger plants bring down the cost significantly to offer a competitive supply. But the higher investment need and the risks related to monopoly supply of husk from the rice mill, organizational challenges of managing a larger distribution area and the risk of plant failure can adversely affect the investor interest. Moreover, the regulatory uncertainties and the potential for grid extension can hinder business activities in this area.

INTRODUCTION

Out of 737 million tonnes of paddy produced in the world in 2012, 90% came from Asia and South Asia contributed about 30% of this production [1]. Rice is cultivated on 60 million hectares in the region and offers livelihood for more than 50 million households [2]. Rice cultivation and processing (i.e. milling) thus form major economic activities in the rural areas of the region. As a by-product of rice production, the region also produces a significant amount of straw (about 225 Mt y^{-1}, assuming a straw to paddy ratio of 1.0), and rice husk (about 45 Mt y^{-1}, assuming a paddy to husk ratio of 0.2) but the residue or waste does not produce any commercial value in most places. A part of the waste is used as fodder (or in animal food preparation) and in brick kilns, while a significant part of it is used as a source of energy, mainly for cooking purposes and for parboiling of rice. The rest is burnt in the field, creating environmental pollution.

However, few attempts have been made in South Asia, particularly in India, to utilize rice husk for electricity generation. The Ministry of New and Renewable Energy in India has been promoting biomass gasification projects under various schemes and it is reported that there are 60 mini rice husk powered electricity plants operating in various parts of the country. But the success of Husk Power Systems (HPS) as a private, off-grid electricity producer and supplier has renewed the commercial interest in this waste-to-electricity conversion. Other countries in the region however have been slow in exploiting the resource commercially. There is a single 250 kW rice-husk based power plant operating in Bangladesh (namely Dreams Power Limited in Gazipur), while Pakistan does not seem to have yet experimented with rice husk as a source of power. The lessons from successful commercial ventures in the region can support wider application of this waste to electricity technology.

Yet, the issue of rice husk-based power in South Asia has not been widely analyzed so far. There are few case studies discussing specific projects or initiatives but there is a dearth of systematic academic studies. The purpose of this paper is to analyze the business model and techno-economic feasibility of rice husk based electricity generation to understand the basic conditions required for developing a viable husk power business. This study thus intends to bridge the knowledge gap indicated above.

The paper is organized as follows: Section 2 presents a review of the HPS business model and DESI Power model; Section 3 considers rice-husk based electricity generation for rural applications with and without rice mill demand and checks for the viability of such a system. It then expands the system size to consider the viability of operation at the village cluster level. Finally some concluding remarks are presented in the last section.

EXPERIENCE OF USING RICE HUSK FOR POWER GENERATION

Rice husk is being used for electricity generation in India in various sizes of plants. Rice husk gasification is commonly used in small-scale plants for electricity access and two commercial ventures, namely the HPS and DESI Power have been quite successful in this respect. This section presents these experiences.

The HPS Model

The Husk Power Systems offers an example of an innovative rural electrification business that has combined electrification with rural development by providing access to electricity while ensuring environment protection, wellbeing of local population and empowerment of local communities. The "rural empowerment enterprise", headquartered in Patna, Bihar, was set up in 2007 by a group of like-minded friends who wanted to challenge the conventional perception of rural electrification as a non-viable business proposition. They realized that the low cost electricity supply and high quality service are key factors for a viable, small-scale electricity generation option. HPS looked into various alternative electricity generation options (such as wind turbines, solar photovoltaic (PV), biodiesel, and bio-gas) but selected rice husk, an abundant local resource in the rice-growing region of the country which was hitherto treated as a waste, that can be procured at very low cost for conversion to electricity.

HPS decided to use the gasification technology which is a partial oxidation process in which a solid fuel is mixed with oxygen (air) in a controlled system to produce producer gas. However, husk does not burn easily due to its high silica content and silica-cellulose structure that causes wear and tear of components coming in contact with it [3]. Although biomass gasifiers have been used previously in India, most common applications involved dual fuels where diesel is used as a support fuel but this increases the cost of electricity generation. For a mono-fuel application, a customized, proprietary design of gasification technology was required that can be built and maintained locally without high level technical expertise. The mono-fuel gasifier was locally fabricated and a cheap compressed natural gas (CNG) engine was procured to start the initiative in 2007. The decision of local fabrication of the equipment as opposed to buying from a manufacturer has turned out to be a smart move as this has reduced the capital cost of the plant.

In addition, the company has been relying on smart technologies to reduce operating costs and potential revenue losses. The distribution mini grid uses smart features for remote monitoring of the system. The company uses smart meters for billing purposes and the bill collectors use hand-held data recorders to keep record of the collection made from door-to-door bill collection rounds. The use of insulated cables hoisted on bamboo poles reduces the potential for electricity theft while reducing the capital investment required for the distribution network.

To ensure regular supply of feedstock, the power plants are located in the paddy growing area and more importantly, close to rice mills to create a win–win situation for both. The mill provides a steady supply of husk and offers a base load demand for the power plant that helps achieve a better plant utilization rate for the power plant, which in turn reduces the average cost of supply. The power plant supplies electricity that is reliable and cheaper than the alternative sources like diesel generators. In some cases, the company has integrated the rice mill business to ensure business viability [4] and to internalize the symbiotic relationship with the power plant.

In addition to careful plant siting, HPS has also taken advantage of other income generating opportunities and created a community support system to ensure better integration with the community. It has registered a Programme of Activities (PoA) under the Clean Development Mechanism for its electrification activity that aims at providing electricity to non-electrified areas through renewable energy sources. The PoA will remain active for 28 years (until 2040) and the emission reduction has been sold in advance to generate carbon revenue. Further, the char obtained from burning the husk is used for incense stick making, thereby monetizing the waste. Local women are employed in incense stick-making, thereby reinforcing the development link with the community. It is reported in [7] that a 32 kW plant produces 6 t of incense sticks per year. Silica precipitation is sold for mixing with cement. The innovative approach towards revenue generation from various products surely helps in improving its financial position.

The HPS claims that in the process it returns more to the local community than that it collects through its electricity bills. Each plant engages 3–4 staff – a plant operator, an electrician, a husk loader and a bill collector, who are taken from local youths and trained by the company, through which about 400 $ per month is recycled into the local economy. The rice mills supplying husk to the power plant receive about 25 $ per tonne of husk (or about 2500 $ per year for a 32 kW plant), an extra source of income for the mills that is often shared the rice mill customers through a reduced fee for milling. The incense stick making activity mentioned previously also provides earning opportunities to local women. In addition, in some cases, the bill collector also acts a "travelling salesman" who takes orders from the households, procures them in bulk from the nearby town and delivers to the households for a small commission. This ensures an extra income for the bill collector and the households get their goods at wholesale rates. This inclusive business model (see Fig. 1) has worked well for the company.

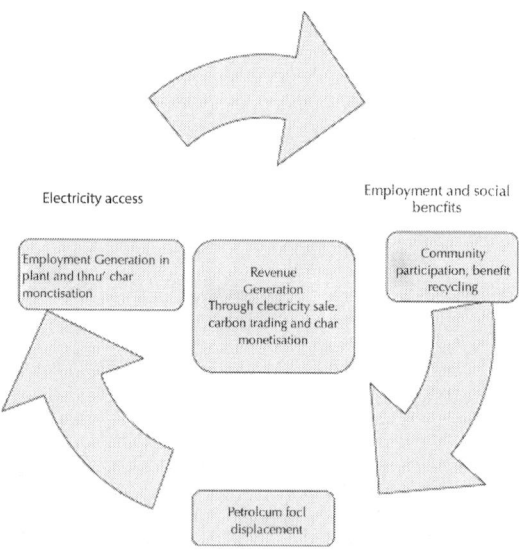

Figure 1: The HPS business model.

The company has successfully extended its business to more than 300 villages to provide electricity to more than 200,000 people installing 84 plants. HPS initiates the process for installing a new plant upon receipt of a request from a village or the local authority, for which an initial deposit is taken from the interested villagers to cover up to three months cost of electricity. Upon enlisting the interest of sufficient number of consumers, the feasibility of a biomass-based plant is carried out, which identifies a secure source of fuel supply for the plant, and verifies the economic viability of the business. The installation process takes about three months and a local team is set up to operate the system on a daily basis. A typical plant can serve, depending on the size of the village and willing consumers, up to 4 villages with about 400 consumers within a radius of 1.5 km of the plant.

The supply is given for a fixed period of time, normally for 6–7 h in the evening using a 3 phase 220 V system. Consumers pay a connection charge and a flat monthly fee (varying between 2 \$ and 2.5 \$) for the basic level of service (2 compact fluorescent lamps and a mobile charging point, called the 30 W package). However, customized packages are also available and consumers with a higher level of consumption benefit from a lower unit rate. Small commercial enterprises are also supplied with electricity but they generally pay a higher flat rate of 4–4.5 \$ per month due to higher demand.

The HPS aims to provide electricity to 10 million people in 10,000 villages by installing 3000 plants by 2017. It has successfully managed to secure funds from a variety of sources in the past, including charitable sources and financial institutions. Although the plants initially followed the Build, Own, Operate and Maintain (BOOM) model, the HPS is also employing other modes of operation, namely the Build, Own and Maintain (BOM) and Build and Maintain (BM) lately to grow faster. In the BOOM model, the company looks after the entire chain of the business, which in turn requires a dedicated set of staff that needs growing with new plants. The overhead can be high and the company faces the investment challenge. In the BOM model, the business is partly shared with an entrepreneur who makes a small contribution to capital (about 10%). The HPS maintains the plant and gets a rental fee but the operational aspects are taken care of by the entrepreneur. This reduces some of the management tasks for the HPS, and builds a local network of entrepreneurs but the HPS still faces investment challenge. Moreover, verifying the quality of the local entrepreneur is a challenging task and the speed of replication using this approach remains unclear. The company transfers the ownership after a specified period of time, upon recovering the cost of investment. The Build and Maintain model essentially transforms the HPS into a technology supplier where its role is limited to supply of the equipment for a fee and maintaining the plant through a maintenance contract. The supply business is undertaken by a local entrepreneur and the HPS does not get involved in this activity, although the entrepreneur uses the HPS brand for the supply. Thus the business uses the franchisee model in this case and as long as the franchisee is able to finance the investment and is capable of running it effectively, the business can grow. Although this is a proven approach in many other businesses, in the context of rural electricity supply this has not been widely used. This model requires a strong quality control and standardization of the business operation but it is not clear whether or to what extent this has been developed in HPS.

Thus, a rapid replication of activities which is necessary for achieving the company target of electrifying 10 million people by 2017 depends to a large extent how the above business models work. This expansion demands significant energy resources, financial resources, management capabilities, skilled local staff, and commensurate manufacturing capabilities. It is not clear whether the company can ensure all the success factors to ensure a rapid growth. It is reported that the husk price has significantly increased since its plants have started operation. Moreover, the niche areas for its operation where consumers can afford high tariffs may be difficult to find in the future, which can limit the growth prospect. Critics point out that HPS only operates in niche areas where villages had been receiving diesel-based electricity from local entrepreneurs and the relatively rich consumers in those areas were already paying high charges for their electricity. HPS has thus displaced diesel-based generation by offering electricity at a cheaper rate. In addition, the plant size ranges between 5 kW and 250 kW, which fails to exploit the economies of scale and scope and affects the business prospects. Clearly, the replication issue requires further investigation

DESI Power

The Decentralised Energy Systems India Private Limited (DESI Power), a not-for-profit company set up in 1996 with aims to provide affordable and clean decentralized energy to rural communities for rural development, offers an integrated solution comprising of building and operating decentralized power plants, creating rural service infrastructure through mini/micro grids, engaging with the local community for establishing partnership models and organization structures for community-based management of the services, and providing training for capacity building in rural areas for micro-enterprise and business development. These activities are undertaken through sister organizations, such as DESI Power Gramudyog (DPG) for village level businesses and enterprises and DESI Mantra for training and capacity building. In addition, joint ventures and partnerships are also established for energy service and village enterprises.

The first plant of DESI Power was set up in 1996 in a village in Madhya Pradesh (India) and relied on biomass gasification systems. It has set up 16 power plants in total by 2012, with installed capacities ranging between 11 kW and 120 kW. In most cases, DESI Power acts as the rural independent power producer and enters into a power purchase agreement with the buyers' organization (which could be an individual entity, a co-operative society or an association of buyers). It serves mainly rural enterprises and small industries that would otherwise rely on diesel generators for their electricity supply to complement unreliable grid supply. It also assists in the development of micro-enterprises, often linked to agriculture. The company also enters into biomass purchase agreements with local suppliers (who can be villager groups or commercial suppliers).

However, beyond this niche area of operation, DESI Power has also installed four mini-grid systems to supply electricity to households, micro-enterprises and mobile phone towers, where an anchor load (such as the mobile phone towers) is generally included in the system that offers the base load and increases the financial security for the operation. Until 2012, 10 mobile phone towers have been connected to its existing power plants and it plans to expand this to another 20 towers in two years. Moreover, the emphasis is on generating as much electricity as possible through the inclusion of micro-enterprises. This reduces the average cost of supply that in turn enhances viability of the micro-enterprises. This inter-dependence is exploited to ensure affordable power as well as rural economic development.

In addition to focussing on the niche market, there are other distinctive features of DESI Power approach to the business. DESI Power has installed underground cables to connect consumers, which is a costlier option, although it is less prone to theft and is a more secure option. Its pricing policy is based on the services it offers and not often focuses on electricity pricing as such. For example, for a light point of 60 W a fixed rate of 8.3 US cents (or 5 Indian rupees) per day is charged while micro-enterprises pay a fixed fee for the service. Similarly, one hour of irrigation water supply from a 3.75 kW pump is charged at 1 $ (or 60 Indian rupees) [5]. The company also offers a range of bill collection options – daily for small households and micro-enterprises and monthly for bigger industrial/institutional consumers. Although this appears to be working for them at the moment, the daily collection of revenue is a labour intensive, costly option. Moreover, it follows the Build-Operate-Transfer model of operation wherein it hands over the plant to the local community or village groups after a period of operation.

Like HPS, DESI Power has also registered a small-scale project with the CDM Board for establishing 100 biomass gasifier-based decentralized, power plants in the District of Araria in Bihar state (India). The plants will be of 50 kW capacity with the exception of a few 100 kW plants. In total, 5.15 MW of capacity was expected to be installed by 2012 which will reduce about 360 k tonnes of CO_2 emission over the first ten years of the project. However, with only a few plants set up so far, the company has significantly underachieved in terms of emissions reduction and capacity addition targets. Although the company expansion plan maintains that it aims to achieve its 100 village target in 3 to 5 years, and would establish 5 pilot plants in 2013, the outlook remains uncertain.

Apparently, the investment challenge is the most important barrier faced by DESI Power. While a 50 kW gasifier plant costs 45,000 $, an equivalent diesel generator capacity costs just 10,000 $. In addition, the village co-operatives or associations have limited borrowing capacity and do not have the required deposit or bank guarantees for availing any debt finance. Similarly, in the absence of a bankable agreement with the co-operatives or the buyers, the company cannot finance its projects. This constraint appears to be having a significant effect on the business expansion of the company. In addition, the technical capacity to deliver plants and human capacity to operate and maintain them are also constrained.

BUSINESS CASE OF POWER GENERATION FROM HUSK

To analyze the economic and financial viability of rice-husk based power supply business, we present a set of cases based on available information and realistic assumptions. The proprietary nature of financial information of existing companies leaves some information gap. Further, some costs, particularly the capital cost of biomass gasifiers can vary depending on the technology source, components used and the degree of environmental protection considered at the project site. The analysis presented here follows a scenario approach where different plant sizes, different levels of electricity demand and alternative capital structures are considered to develop a better appreciation of the range of business outcomes. First, a 20 kW plant is considered, which is followed by a case of 200 kW plant.

Providing Electricity Access to Households with a 20 Kw Plant

In this case, the plant serves about 400 households, most of whom may be consuming a minimum amount of electricity for lighting and mobile phone charging. However, to allow for different consumption behaviour, particularly by those who can afford to consume more, alternative demand scenarios are developed inTable 1.

Table 1: Alternative demand scenarios

Description	Scenario 1	Scenario 2	Scenario 3
Total number of households serviced	400	400	400
% of HH using basic 30 W service	100%	90%	85%
% of HH using a medium level of 75 W	0%	10%	10%
% of HH using high demand of 250 W	0%	0%	5%
Number of commercial units	20	30	30
Demand by commercial units	60 W	60 W	60 W
Hours of service	6	6	6
Days of operation	365	365	365
Electricity demand (kWh per year)	28,908	34,164	43,800
Required Plant utilization (for a 20 kW plant)	16.5%	19.5%	25%
Plant loading (for 6 h of operation)	66%	78%	100%

In scenario 1, all 400 households use the basic level of electricity and there are 20 small commercial units, each with a load of 60 W. A 20 kW plant can easily meet the demand just using only 16.5% of the capacity (or 66% of the capacity considering a 6 h cycle). The second scenario allows for differential household demand based on consumer mix. It is assumed that 90% of the households use the basic level of demand while the rest 10% use a moderate level of electricity at 75 W per household. In addition, 30 commercial units are considered instead of 20 units each using 60 W. The demand increases marginally but a 20 kW plant still can service the load at 78% loading for a 6 h cycle. The third scenario modifies the residential load slightly to ensure a 100% loading of the plant. Yet, as the plant runs for a fixed period of 6 hours, its overall capacity utilization does not exceed 25%. This is relatively low for a power generating plant.

The financial analysis of the 20 kW generator plant is based on the following assumptions:

- The capital cost per kW of installed capacity is taken as 1300 $. The Indian companies like HPS or DESI Power have reported a much lower cost (800 $ kW^{-1}) but HPS manufactures its gasifiers locally using its own design while DESI Power procures its technology from Netpro, promoted by DASAG, which is also a stakeholder of DESI Power. Such special cost advantages may not be available to other projects, which justify a higher cost used here. It is important to mention that a recent report [10] has suggested much higher costs based on a project in Cambodia.

- The cost of distribution network per kilometre is taken as 2000 $ km^{-1}. This cost can vary depending on the quality of the network, materials used, terrain, and cost of labour. For underground cables, the cost may be higher while for distribution systems using bamboo poles, it may be lower.

- The monthly operating cost is considered to be about 100 $ [6] and [7] – or about 4% of the capital cost.

- Each plant employs four employees with a salary of 100 $ per month on average, which is close to the monthly salary cost of 380 $ indicated in [7].

- The plant life is taken as 15 years and the plant operates 6 h per day, every day of the year.

- Husk price is taken as 25 $ per tonne and it is assumed that the fuel price remains unchanged over the project lifetime.

- The lower calorific value of husk on dry basis is taken as 12.6 MJ/ kg [3] and [4] and the conversion efficiency of gasifier is taken as 20%.

- The cost of debt is taken at 5.5% y^{-1} while the rate of return on equity is taken as 10% y^{-1}. The weighted average cost of capital is used to determine the discount rate.

- A straight-line depreciation is used after allowing a 10% salvage value for the asset at the end of its life. Where grant capital is used, it is assumed that the grant capital reduces the capital required for investment and the depreciation charge is reduced accordingly. Although the grant capital can be treated differently in accounting terms, the above provides a simple treatment of the grant.

- It is assumed that the company is not paying any tax and hence the tax benefit arising from debt capital does not apply here.

The cost of electricity supply for different scenarios and considering alternative debt-equity combinations and grant capital share is calculated. The levelised cost of electricity supply is used as the indicator. The levelised cost is the real, constant cost of supplying electricity that if recovered from consumers over the lifetime of the plant would meet all costs associated with construction, operation and decommissioning of a generating plant. This generally considers capital expenditures, operating and maintenance costs, fuel costs, and any costs involved in dismantling and decommissioning the plant. Equation 1 provides the mathematical relationship for the levelised cost of electricity.

$$LCOE = \frac{Present\ value\ of\ (capitel\ cost + O\&M\ cost + fuel\ cost)}{Present\ worth\ equivalant\ of\ (electricity\ consumed)} \quad (1)$$

For scenario 1, the result of the levelised cost analysis is shown in Fig. 2. As expected, the lowest levelised cost is obtained when the entire capital requirement comes from grants and the cost for this scenario comes to 270 $ MWh^{-1}. But if no grant is received, the cost of supply that has to be borne by the consumers varies between 400 $ MWh^{-1} to 490 $ MWh^{-1} depending on the share of debt and equity. This clearly shows that part load operation of the system is a costly option despite the low capital cost per kW compared to other technologies (such as solar PV or wind). Clearly, both HPS and DESI Power have realized this and used adequate households and/or micro-enterprises to ensure high plant capacity utilization.

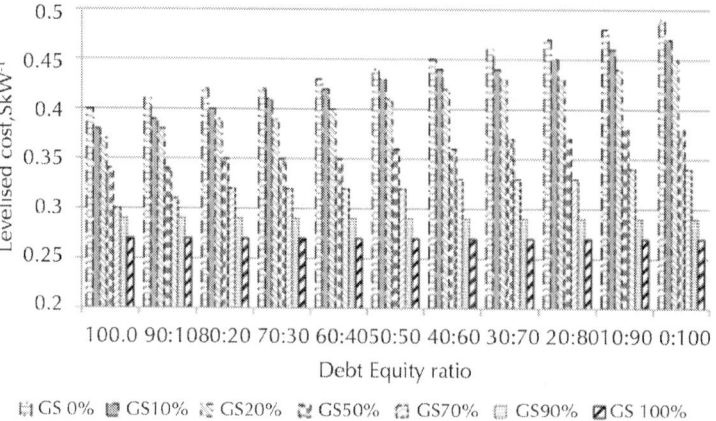

Figure 2: Levelised cost of electricity supply for scenario 1. Note: GS – grant share.

However, the important issue is whether or not a flat rate charge of 2 $ or 2.5 $ per month per household can recover the expenses in scenario 1. As the consumers use only 5.5 kWh per month, their effective tariff varies between 0.36 $ kWh^{-1} and 0.46 $ per kWh depending on 2 $ and 2.5 $ monthly charges, which is considerably higher than the prevailing rate for grid-based electricity. Therefore, as long as the levelised cost of electricity supply is below the above tariff, the business becomes viable in this scenario. If the company charges 2.5 $ per month, even without subsidy it can operate the business profitably as long as the debt –equity ratio is not worse than 50:50. If it charges 2 $ per month, the company needs at least 50% capital grant subsidy to run the business, unless other sources of income can make up for the loss. As other income tends to be limited in nature, it becomes clear that providing access to poor households with limited demand remains a vulnerable business.

In scenario 2, the levelised cost of supply reduces to 0.24 $ per kWh for a capital subsidy of 100% while the cost varies between 0.34 $ and 0.42 $ per kWh for no capital subsidy (see Fig. 3). Although better plant utilization reduces the levelised cost of supply, the revenue would not change if all residential consumers are charged at the same rate. Consequently, when different consumer categories use different levels of electricity, a differential tariff is required to recover the cost. A higher flat rate for the high-end consumers constitutes a simple solution in this case, which may end up in a lower average rate for this category due to higher electricity consumption. The tariff per Watt instead of watt-hours is thus a simple but effective way of passing higher charges to poorer consumers in disguise.

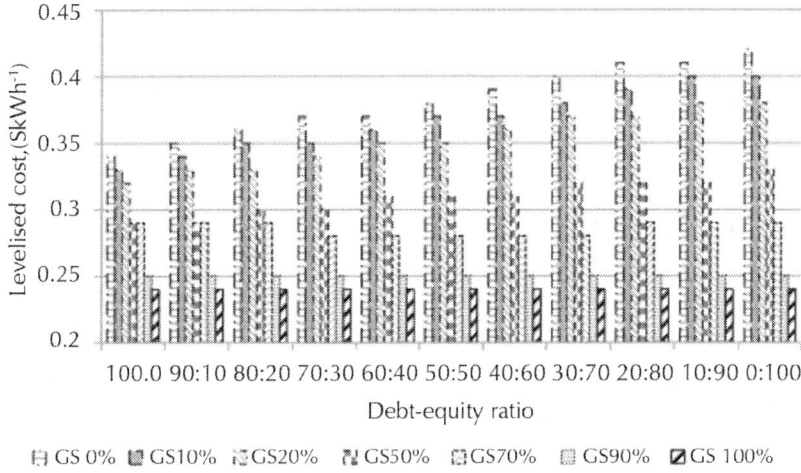

Figure 3: Levelised cost of supply for scenario 2. Note: GS – grant share.

In the third scenario where the capacity is fully utilized for the 6 h period of supply, the levelised cost reduces even further. This scenario, as expected, produces the lowest cost of supply (see Fig. 4) and the levelised cost with full grant funding compares quite well with the low rates reported by HPS. It needs to be mentioned that this analysis used a higher capital investment cost compared to that reported by HPS, which excludes the possibility of achieving the same outcomes as HPS. The result of this scenario supports the claim made by HPS that they are in an advantageous position compared to other renewable technologies. Yet, the levelised cost of supply still remains quite high compared to the grid-based supply in the absence of any capital subsidy. However, the tariff for grid-based supply may not be a true comparator given the unreliable and poor quality of supply. Consumers tend to spend considerably higher amounts for alternative sources of supply (e.g. from generator sets). Accordingly, the willingness to pay for a reliable supply is likely to be higher than the tariff for grid supply, particularly for commercial and industrial consumers.

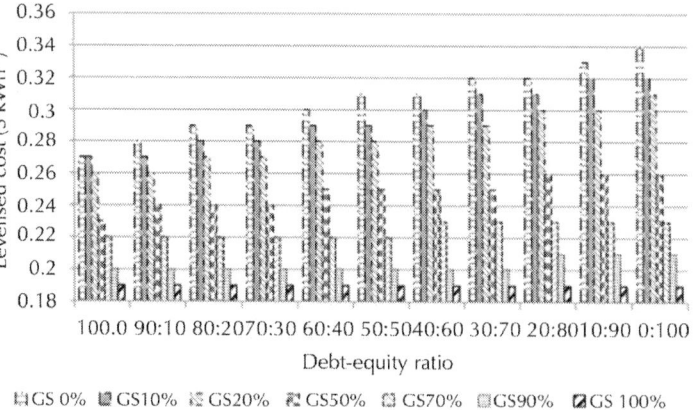

Figure 4: Levelised cost of electricity supply for scenario 3. Note: GS – grant share.

Once again, a differential tariff will be required to ensure adequate revenue generation. From the company's perspective, running the plant near its full load will ensure higher profitability and clearly, this will ensure that the operation can be sustained with limited or no financial support. But grant capital surely contributes towards risk mitigation and acts as an incentive for the supplier.

It can be concluded that limited electricity supply for a fixed duration can be effectively provided using the rice-husk based system. The cost-effective operation of the system however requires careful control of two factors: 1) that the plant capital cost needs to be minimized, perhaps through local and indigenous design of the plant; and 2) the plant is operated near full load by enlisting adequate number of consumers, preferably with some demanding more than just the basic level of supply (30 W per household). Although the cost of supply remains higher than the prevailing grid-based supply, the business can be run viably with a suitably designed tariff system. The difference in the approach between HPS and DESI Power can be understood from this analysis. HPS has ensured viability by enlisting adequate number of residential customers whereas DESI Power enlisted the support of micro-enterprises. This avoids reliance on a large number of very small consumers as the business or commercial load tends to be much higher than the basic level of residential demand. However, the cost per kWh incident on the poor tends to be higher than those consuming more in the absence of any cross-subsidy or direct subsidy. This tends to be true in any electricity system – more so in a privately owned and operated system, but mitigating measures are often used through direct social safety nets and/or subsidized supply schemes. Hence any support for additional income generation will surely be beneficial.

Electricity Access with Rice Mill as an Anchor Load

Given that the electricity plant in the previous case has idle capacity outside the evening peak hours, the power plant could consider adding new demand to improve its financial viability. The rice mill offers such a load: it may not have a good quality power supply and the cost of supply may be much higher than the rice-husk based supply. This alternative case is considered below.

Clearly, the energy demand for a rice mill will depend on its size, processing activities involved, level of automation, operation time and such factors. For the purpose of this analysis, the following assumptions are made:

- The rice mill capacity is chosen in such a way that adequate husk can be sourced from the mill to meet the demand for electricity generation;
- Small mills in a village or small town location tend to be indigenously made and tend to consume more energy. It is assumed that the electricity consumption requirement per ton of raw rice processed is 43 kWh Mg^{-1} [3].
- Rice mills in India can be categorized into two broad groups: small sized ones with less than 1 tonne per hour processing capacity and bigger mills. Small mills generally operate a single shift of 6–7 hours for about 200 days (i.e. 1200 h of annual operation) while larger mills run two shifts (between 2400 and 3000 h of annual operation). In this case, we assume a single shift operation for 1200 h per year.
- Husk availability is estimated considering a husk to paddy (or raw rice) ratio of 0.2.
- It is assumed that the rice mill operates during day time when the residential demand is not serviced. This in effect extends the hours

of operation of the power plant. Electricity demand is unlikely to be constant for the entire period of operation. It is likely that the evening load may be higher than the day load. For the sake of simplicity of financial analysis, an equivalent plant loading is used that generates the total amount of electricity required to meet the total demand.

- Scenario 3 from the previous section is used for electricity demand for non-mill purposes.
- The power plant operates two shifts of 6 h and instead of 4 employees uses 6 employees, each receiving a monthly wage of 100 $. This is logical given that the work for bill collector and the plant technician does not increase proportionately with hours of plant operation.

The rice mill has to be such that it produces enough rice husks in a year to meet the electricity needs of the mill and the village community. Given that the electricity demand corresponding to scenario 3 is 43.8 MWh, and considering 43 kWh electricity required for processing one tonne of rice, we find that a rice mill of 0.4 t h^{-1} capacity operating in a single shift of 6 h for 200 days in a year will produce sufficient rice husk. The rice mill will require 20.64 MWh of electricity and the power plant needs to produce at least 64.44 MWh per year. The rice mill will process 480 tonnes of raw rice per year and will produce 96 tonnes of husks per year. The power plant will require approximately 93 tonnes of husks for its operation, which can be procured from the rice mill directly.

Fig. 5 presents the levelised cost of supply for the integrated power supply operation to the rice mill and the village community. As can be seen, the cost of supply reduces considerably in this case due to higher plant utilization rate. The lower end prices with capital subsidy will be quite attractive to most consumers. Even otherwise, the cost of supply reduces significantly. Hence, it makes economic sense to extend the supply to the rice mills, particularly when the operation does not coincide with the peak demand. This will benefit the rice mill by reducing its dependence on grid electricity, and providing a reliable supply at a reasonable price. Other consumers also benefit from this integration as the overall cost of supply reduces.

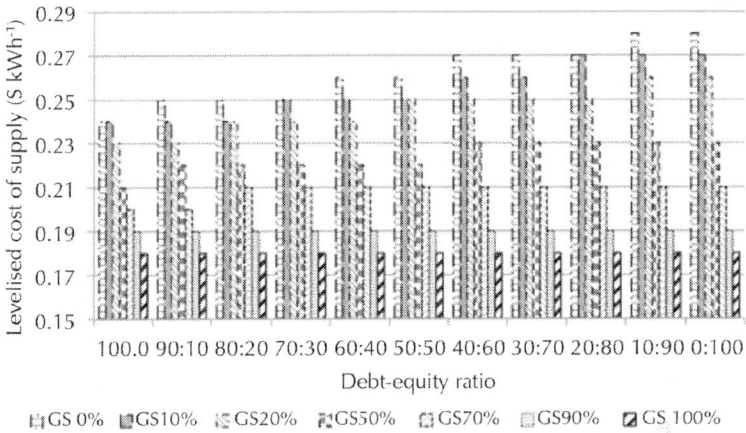

Figure 5: Levelised cost of electricity supply for integrated operation. Note: GS – grant share.

Although rice mills can install power generating stations for own use, such installations may not qualify for government support schemes for rural electricity supply. Moreover, the skill requirement is very different for operating a power plant and electricity distribution business compared to running a rice mill. In organizational terms, it makes better sense to have separate entities dealing with two separate businesses but linked to each other through contracts for fuel supply and electricity supply. Such contractual arrangements are important to ensure risk sharing, bankability of investments and reliability of business operations. The captive power supply model used by DESI Power follows this example.

VIABILITY OF A SCALED-UP ELECTRICITY SUPPLY SYSTEM

Rice is the staple food of 1.7 billion people in South Asia. Rice is cultivated on 60 million hectares of land in the region and about 225 million tonnes of rice are produced annually, contributing 32% of global raw rice production. Five major rice producers in the region are India, Bangladesh, Pakistan, Nepal and Sri Lanka. India produced about 158 Mt of raw rice (or paddy) in 2012 [1] and has a total rice milling capacity of about 200 Mt per year [8]. In addition, Bangladesh produced 51 million tonnes of paddy in 2012 while Sri Lanka, Nepal and Pakistan produced about 17 Mt of paddy [1].

Rice milling takes place both at the household level (using hand pounding or pedal operated systems) and in rice mills. Generally, a small amount of raw rice is processed at the household level, mostly for own consumption. The processing of raw rice takes two forms: dry hulling which tends to account for a small share of total paddy processing and processing of parboiled rice. Rice milling in the region was a licensed activity for a long time that reserved the activity to small and medium-scale industries. This resulted in the proliferation of small mills throughout the region. However, these mills tend to be inefficient and produce poor quality output (higher percentage of broken rice). Moreover, because many of them fall under the unorganized sector, there is no systematic information about the number, distribution and size of rice mills. However, it is generally believed that the mini mills can process 250–300 kg of paddy per hour, small mills have a capacity of 1 tonne per hour whereas larger, modern mills have capacities ranging from 2 tonnes per hour to 10 tonnes per hour. Smaller mills operate a single shift of 6 h while modern mills operate 2 shifts or even 3 shifts but tend to have a seasonal operation.

Assuming a 2-shift operation of modern rice mills for 200 days per year, and considering that about 30% of the electricity that can be produced from the husk can be used to meet the energy needs of the mill, a simple estimation is made of potential excess electricity and the potential number of consumers that can be served to meet the basic demand of 30 W per consumer for 6 h a day for every day of the year (seeTable 2). It can be seen that thousands of consumers can be served by such power plants and a large cluster of villages (or blocks) can be considered as the basic unit of electrification. Alternatively, excess electricity from the mills can also be sold to the grid if mills are grid connected or can be sold to a small number of local productive users (e.g. irrigation pumps, flour mills, food storage, etc.). Such larger plants thus open up the possibility of including productive applications of electricity beyond rice mill use, which in turn can catalyze economic activities at the village level. Although agriculture is the main rural activity in South Asia, food processing and other agro-based industrial activities (such as storing and warehousing), play a limited role yet due to lack of infrastructure and reliable electricity supply. While small-scale generating plants can only provide limited supply to households and small commercial consumers, larger plants can act as an agent for rural development.

Table 2: Potential for serving large consumer bases

Mill capacity (t h⁻¹)	Husk production (t y⁻¹)	Potential electricity output (MWh y⁻¹)	Mill consumption (MWh y⁻¹)	Excess electricity (MWh y⁻¹)	Number of basic demand consumers that can be served
2	960	672	206.4	465.6	7087
3	1440	1008	309.6	698.4	10,630
4	1920	1344	412.8	931.2	14,174
5	2400	1680	516	1164	17,717
6	2880	2106	619.2	1486.8	22,630
8	3840	2688	825.6	1862.4	28,347
10	4800	3360	1032	2328	35,434

In terms of cost of supply, two opposing forces are expected to operate. On one hand, the unit cost of generating plant ($ MW⁻¹) is likely to reduce as the size increases. On the other, the fuel cost, distribution cost and wages would increase. The fuel cost increases proportionately with power generation. The area to be served may increase disproportionately and the extension of low voltage lines over long distances will increase distribution losses and affect power quality. This will require a distribution system at 11 kV or higher voltage level and accordingly, the cost will increase. Finally, the staff requirement will increase in proportion with the area being serviced. Billing and collection cost can increase rapidly. Accordingly, the accurate cost estimation is rather difficult in this case.

To obtain a rough idea about the economic viability of a larger plant, the following assumptions are retained:

- A rice mill of 2 t h⁻¹ is considered. This can feed an electricity plant of 200 kW.
- The capital requirement per kW to be 1000 $ for a 200 kW plant;

- 25 staff will be employed for generation, distribution and supply management;
- The distribution system is extended over a distance of 20 km;
- Other assumptions remain unchanged. It is possible to consider 24 h operation of the power plant but in this case, the available rice husk can support a smaller power plant capacity. Moreover, a husk-based plant is unlikely to operate continuously for 24 h. In this case, a back-up will be required. For these reasons, a two-shift operation is retained here.

The levelised cost of electricity for no subsidy case comes to 190 $ MWh^{-1}. The cost reduces further with different levels of subsidy (see Fig. 6). The levelised cost in this case is the lowest of all options considered in this study. Clearly, this shows that as long as sufficient number of willing consumers can be enlisted, and the power supply company can manage to run its village cluster level operations, a bigger business can be profitably run. Alternatively, the excess power can be sold to captive users or to the grid at a break-even price of 190 $ MWh^{-1} to make the venture viable. However, the tariff offered by the utility for buy-back is not as remunerative as this, which hinders financial viability of such power plants.

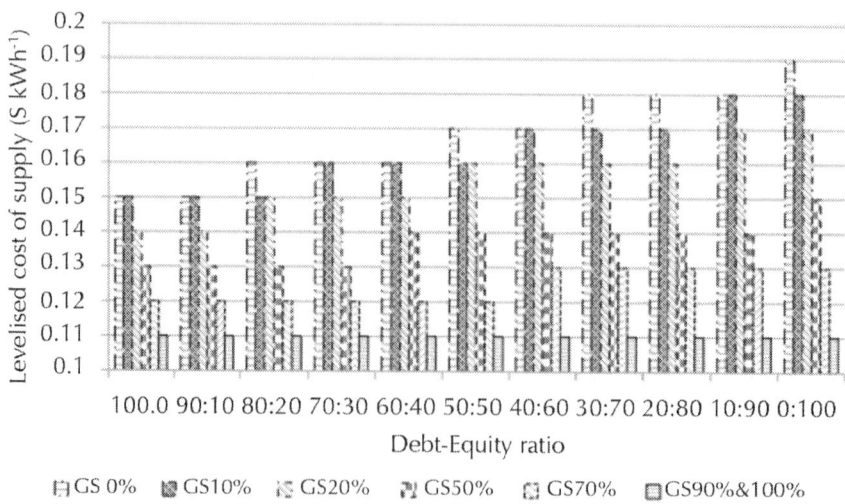

Figure 6: Levelised cost of electricity supply for a 200 kW plant. Note: GS – grant share.

Clearly, such a scaling-up of the business has its pros and cons. A bigger plant and larger area of operation may be more attractive to investors willing to enter in the mini-grid business. Such plants offer some economies of scale and therefore can be a more efficient option economically. It may also be possible to take advantage of carbon credits either through the CDM programme or through other voluntary offset schemes. The byproducts of electricity production and the symbiotic relationship between the rice mill, power plant and the local community can support such systems positively.

Yet, the risks involved in such an integrated operation cannot be overlooked either. The power plant is heavily dependent on the rice mill and any break-down in the mill or its closure will jeopardize the power plant operation. A 200 kW plant would require about 1000 tonnes of husks per year for its operation and procuring such a volume of husk from an alternative source can be difficult. The transportation cost of feedstock can easily increase the fuel supply cost and render the electricity supply less cost effective. Similarly, if the mill does not settle its electricity bills regularly or delays in making the payments, the bad debt can increase and the cost of running the business can increase. Diversifying the commercial demand can mitigate the over-dependence on the rice mill.

Similarly, as the plant serves a larger area, any fault with the generating plant will result in a power supply disruption in the entire area. While installing two 100 kW plants can be a better option, the capital cost may increase. In addition, the plant would need regular maintenance on a daily basis to ensure proper cleaning of the gas filters and this makes it difficult to run the system continuously. It is likely that a back-up system will be required to meet the essential demand for a limited period of time. Depending on the fuel or technology used, the back-up system can increase the overall cost of electricity supply. Similarly, the distribution network would require greater attention and any fault in the distribution system can reduce the system reliability.

The investment requirement for a plant of 200 kW can easily reach 250,000 $. This is a substantial investment in a rural location, and companies willing to enter into such businesses will need to muster adequate financial resources and relevant experience. Securing long-term debt funds from the financial institutions can be a major challenge as many of them require more than 100% guarantee for such loans. In addition, the loan term (period and interest rate) may not be favourable to this type of businesses. Project financing of mini-grids can be challenging due to limited number of bankable contracts with consumers. Any support from the government and international agencies in facilitating finances through credit facilities, grants and guarantees can be helpful.

The investment challenge amplifies due to regulatory uncertainties in the area of off-grid electrification. As indicated in [9], the supply of electricity through a local off-grid network requires conventional regulatory supervision due to the possibility of monopolistic exploitation of the consumers, supply quality concerns and potential disputes between the supplier and the consumers. However, the regulatory arrangement for mini-grids is not quite clear in many South Asian countries. It appears that the rural areas covered by off-grid supply still come under the jurisdiction of the utility providing the central grid-based supply. Any decision to extend the grid subsequent to the installation of the off-grid plant can make the off-grid business unviable and stranded. Thus, the regulatory uncertainty needs urgent consideration.

CONCLUSIONS

This paper has considered off-grid electrification through electricity generated from rice husk in South Asia. The Husk Power Systems has successfully used rice husks to provide decentralized electricity in rural areas of India and has so far installed 84 plants to electrify 300 villages. The success of the HPS can be traced to their choice of technology that is less capital intensive compared to other renewable energy options, their innovative approaches towards system cost reduction (e.g. using temporary structures made of bamboo poles for distribution network, local manufacturing of gasifiers) and additional income generation (e.g. use of carbon offsets and monetization of byproducts), careful tariff design linked to Watts of demand instead of Watt-hours of energy used and careful siting of plants where about 400 customers are willing to pay for the service. DESI Power on the other hand has placed emphasis on productive use of power and used husk-based systems to displace diesel-based electricity supply to micro-enterprises. It has also used anchor loads (such as supply to mobile telephone towers) to improve the financial viability of the business.

The financial analysis of rice husk-based power generation shows that the levelised cost remains high compared to the supply from the centralized grid when just the basic demand (of 30 W) of households is met. This is due to low plant utilization factor but the tariff based on Watts helps generate the required revenue to run the system. As the system utilization improves either due to higher electricity consumption by some or by integration of the supply system to the rice mill, the levelised cost of supply reduces. However, the benefits of such cost reduction are enjoyed by those who consume more when an inverted block tariff system is used. The integration of rice mill's electricity demand brings the costs down considerably due to extended use of the facility during off-peak hours. Such integration can ensure an anchor load and can be beneficial for the electricity supplier. The rice mill on the other hand benefits from a reliable supply at a comparable price and reduces its cost arising out of electricity disruption. While the rice mill can develop a power plant for its own consumption, it is better to allow a specialized, separate entity to deal with the power generation business and develop contractual arrangements for fuel and power supply.

The extension of the analysis to include larger power plants for electricity distribution to a cluster of villages results in the cheapest cost of supply due to realization of economies of scale. The cost of supply in such a case can be very competitive even without any capital grants. This suggests that it makes economic and financial sense for a supply company to extend the business to cover larger areas as long as there are sufficient willing customers and adequate supply of rice husks from rice mills. This also can promote economic activities in rural areas and promote economic development urgently needed to reduce rural poverty. Yet, the regulatory uncertainty, limited access to financial resources and markets, increased complexity of the distribution network (i.e. it may require higher voltage permanent network systems to reduce losses), and higher dependence on a single or limited fuel supply source would have to be carefully considered. Such bigger systems would require careful system design to ensure adequate system reliability, appropriate maintenance and limited line loss in distribution.

Being a major rice producing region, South Asia surely has a significant potential of utilizing a major agro-waste to produce electricity for rural supply and rural development. However, to realize the potential the barriers mentioned above need to be addressed. In addition, the potential for using rice straw alongside rice husk can also be considered for power generation. Similarly, the potential for replication of this business in South Asia needs further analysis.

ACKNOWLEDGEMENTS

The activities reported in this paper are funded by an EPSRC/DfID research grant (EP/G063826/2) from the RCUK Energy Programme. The Energy Programme is a RCUK cross-council initiative led by EPSRC and contributed to by ESRC, NERC, BBSRC and STFC. The views expressed in this report are those of the authors and do not necessarily represent the views of the institutions they are affiliated to or the funding agencies. Usual disclaimers apply.

REFERENCES

1. Food and Agricultural Organisation. Table 1: world paddy production. Rice Market Monitor 2014 Apr;XVII(1):33.

2. Mohanty S. Rice in South Asia. Rice Today 2014; 13 (2):40e1.

3. Kapur T, Kandpal TC, Garg HP. Electricity generation from rice husk in Indian rice mills: potential and financial viability. Biomass Bioenergy 1996; 10 (5-6):393e403.

4. Sookkumnerd C, Ito N, Kito K. Feasibility of husk-fuelled steam engines as prime mover of grid-connected generators under the Thai very small renewable energy power producer (VSPP) program. J Cleaner Prod 2007; 15 (3):266e74.

5. Krithika PR, Palit D. Participatory business models for off-grid electrification. In: Bhattacharyya S, editor. Rural electrification through decentralised off-grid systems in developing countries. London: Springer-Verlag; 2013. pp. 187e225.

6. Singh S. Empowering villages: a comparative analysis of DESI power and husk power systems: small-scale biomass power

generators in India. Rural market insight brief. Chennai: Institute for Financial Management and Research; 2010 [Cited 2014 May 29]; Available from: http://cdf.ifmr.ac.in/wpcontent/uploads/2011/03/Empowering-Villages.pdf.

7.   Synergie pour l'Echange et la Valorisation des Entrepreneurs d'Avenir (SEVEA). Husk power systems e power to empower. France: Les Marches; 2013 [Cited 2014 May 29]; Available from: http://www.seveaasso.org/wa_files/Case_20study_20HPS_20Vcomprime_CC_81.pdf.

8.   National Skill Development Corporation. Human resource and skill requirements in the food processing sector: study on mapping human resource skill gaps in India till 2022 e a report [Internet]. New Delhi: undated; [Cited 2014 May 29]; Available from: www.nsdcindia.org/pdf/food-processings.pdf.

9.   Bhattacharyya SC, Dow SR. Regulatory issues related to offgrid electricity access. In: Bhattacharyya S, editor. Rural electrification through decentralised off-grid systems in developing countries. London: Springer-Verlag; 2013. pp. 271e83.

10. Innovation Energie Developpement. Low carbon mini-grids: Identifying the gaps and building evidence base on low carbon mini-grids. Final report. Francheville [France]: Support Study for DfID; 2013 [Cited 2014 May 29]; Available from: https://www.gov.uk/government/uploads/system/ uploads/attachment_data/file/278021/IED-green-min-gridssupport-study1.pdf.

Citations

CHAPTER 1

Helmut Haberl, Karl-Heinz Erb, Fridolin Krausmann, Alberte Bondeau, Christian Lauk, Christoph Müller, Christoph Plutzar, and Julia K. Steinberger, Global Bioenergy Potentials from Agricultural Land in 2050: Sensitivity to Climate Change, Diets and Yields, doi:10.1016/j.biombioe.2011.04.035.

CHAPTER 2

Oluwakemi A.T. Mafe, Scott M. Davies, John Hancock, and Chenyu Du, Development of an estimation model for the evaluation of the energy requirement of dilute acid pretreatments of biomass, doi:10.1016/j.biombioe.2014.11.024.

CHAPTER 3

CHAPTER 4

CHAPTER 5

CHAPTER 6

CHAPTER 7

CHAPTER 8

M.S. Verlinden, L.S. Breccia, D. Zonal, G. Berhongaray, T. De Groote, M. Camino Serrano, I.A. Janssen, and R. Coleman's,. Net Ecosystem Production and Carbon Balance of a SC Poplar Plantation during Its First Rotation, doi:10.1016/j.biombioe.2013.05.033.

CHAPTER 9

Subhes C. Bhattacharyya, Viability of Off-Grid Electricity Supply Using Ricehusk: A Case Study from South Asia, doi:10.1016/j.biombioe.2014.06.002.

Index